T0262205

Biodiesel Production: Feedstocks and Methods

Biodiesel Production: Feedstocks and Methods

Edited by **Kurt Marcel**

LANRYE
INTERNATIONAL

New Jersey

Published by Clanrye International,
55 Van Reypen Street,
Jersey City, NJ 07306, USA
www.clanryeinternational.com

Biodiesel Production: Feedstocks and Methods
Edited by Kurt Marcel

© 2015 Clanrye International

International Standard Book Number: 978-1-63240-079-6 (Hardback)

This book contains information obtained from authentic and highly regarded sources. Copyright for all individual chapters remain with the respective authors as indicated. A wide variety of references are listed. Permission and sources are indicated; for detailed attributions, please refer to the permissions page. Reasonable efforts have been made to publish reliable data and information, but the authors, editors and publisher cannot assume any responsibility for the validity of all materials or the consequences of their use.

The publisher's policy is to use permanent paper from mills that operate a sustainable forestry policy. Furthermore, the publisher ensures that the text paper and cover boards used have met acceptable environmental accreditation standards.

Trademark Notice: Registered trademark of products or corporate names are used only for explanation and identification without intent to infringe.

Printed in the United States of America.

Contents

             Heteropolyacid for Free Fatty Acids Esterification
             Marcio Jose da Silva, Abiney Lemos Cardoso,
             Fernanda de Lima Menezes, Aline Mendes de Andrade
             and Manuel Gonzalo Hernandez Terrones

Chapter 9   **Progress in Vegetable Oils**
             Enzymatic Transesterification to Biodiesel - Case Study   **171**
             Ana Aurelia Chirvase, Luminita Tcacenco,
             Nicoleta Radu and Irina Lupescu

Chapter 10   **Adsorption in Biodiesel Refining - A Review**   **187**
             Carlos Vera, Mariana Busto, Juan Yori, Gerardo Torres,
             Debora Manuale, Sergio Canavese and Jorge Sepúlveda

Chapter 11   **The Immobilized Lipases in Biodiesel Production**   **219**
             Margarita Stoytcheva, Gisela Montero,
             Lydia Toscano, Velizar Gochev and Benjamin Valdez

             Permissions

             List of Contributors

Preface

This book highlights the latest developments and growing trends in the field of biodiesel production. The chapters deal with biodiesel production methods. It also encompasses issues related to the use of inexpensive inedible raw materials and production of biomass feedstock with properties that may help it to produce biodiesel. Biodiesel is commonly produced by the transesterification process. The several methods for carrying out this technique include common batch process, supercritical processes, ultrasonic methods, and even microwave methods. This book would be a rich source of information to the students, professionals and scientists.

This book unites the global concepts and researches in an organized manner for a comprehensive understanding of the subject. It is a ripe text for all researchers, students, scientists or anyone else who is interested in acquiring a better knowledge of this dynamic field.

I extend my sincere thanks to the contributors for such eloquent research chapters. Finally, I thank my family for being a source of support and help.

Editor

Part 1

Biodiesel Production Methods

Gas-Liquid Process, Thermodynamic Characteristics (19 Blends), Efficiency & Environmental Impacts, SEM Particulate Matter Analysis and On-Road Bus Trial of a Proven NO$_x$ Less Biodiesel

Kandukalpatti Chinnaraj Velappan and Nagarajan Vedaraman
Chemical Engineering Department,
Central Leather Research Institute,
Council of Scientific and Industrial Research,
Adyar, Chennai
India

1. Introduction

Biodiesel has gained worldwide attention as renewable and blending agent with some lower gas emissions, besides a slight increase of NO$_x$ emission (Michael & Robert ,1998)in the exhaust gas compared to the petroleum diesel. Vegetable oils (Srivastava & Prasad,2000; Prasad & Mohan,2003) namley soybean, sunflower, cottonseed and rapeseed have been examined for fuel without/ with a small modification in the engine. A number of problems, mainly high viscosity, are associated with vegetable oils when directly used as fuel in the CI engines (Agarwal , 1998; Sinha & Misra , 1997; Roger & Jaiduk 1985). It is difficult to reduce particulate matter (PM) and oxides of nitrogen (NO$_x$) (Mohamad and et al., 2002) simultaneously owing the trade-off between NO$_x$ and PM. Moreover, methyl esters of vegetable oils are sulphur free and possess good lubricating properties (De-Gang et al., 2005). Depending upon the climate and the soil conditions, different countries looking for different type of vegetable oil (Goering et al., 1982; Fernando et al., 2003; Antolin et al., 2002; Freedam et al., 1986; Noureddini & Zhu, 1997; Mohamad et al., 2002; Yi-Hsu & Shaik, 2005; Sukumar et al., 2005) used for the biodiesel production; soybean oil in US, rapeseed oil in Europe, palm oil in Malaysia and Indonesia, and coconut oil in the Philippine are being considered (Barnwal & Sharma , 2005).

In India, out of more than 125 million tons (Arumugam et al., 2003) of rice production, about 6 million tons of rice bran and 1 million ton of RBOBD is produced annually (Table 1). General characteristics of refined rice bran oil are as follows: sp gr, 0.916 kg/m³; ref index 1.47; Cloud index, 17; iodine value, 99-108; saponification value, 180-190; unsopanifiable matter, 3.5(%); smoke point, 213 °C; and fire point, 352 °C. General properties of vegetable oil based biodiesel (Table 2) show many variations that might be due to the conversion to biodiesel through different raw materials and different processes.

Country	Rice	Rice bran	Oil
China	181	14.5	2.47
India	137	6.8	1.02
Indonesia	50	4.0	0.68
Bangladesh	38	3.0	0.51
Vietnam	32	2.6	0.44
Thailand	24	1.9	0.32
Myanmar	20	1.6	0.27
Philippines	13	1.0	0.17
Japan	11	0.9	0.15
Brazil	10	0.8	0.14

Table 1. Annual production (metric million tons) of rice, rice bran and oil in the world.

This chapter presents process for rice bran oil biodiesel (RBOBD) production, composition and physico-chemical properties of RBOBD, engine test results, scanning electron microscope (SEM) image, particulate matter in exhaust gas and emission reductions.

Oil Name	ρ (kg/m³)	μ (mm²/s)	Cetane No:	Calorific value (MJ/kg)	Flash point (°C)
Pea	0.883	4.9	54	33.6	176
Soya	0.885	4.5	45	33.5	178
Bab	0.875	3.6	63	31.8	117
Palm	0.880	5.7	62	33.5	164
Sun	0.860	4.6	49	33.5	183
Diesel	0.855	3.06	50	43.8	76
B20	0.859	3.2	51	43.2	128
Range	0.85-0.88	3.2-5	45-62	32-44	76-183

Table 2. Properties of Biodiesel from vegetable oils.

2. Process description

Oil having specific gravity in the range of 0.85 – 0.96 and iodine value not exceeding 208 is heated to a temperature not exceeding 120°C for not less than 2hrs to adjust the moiture content at a level not exceeding 0.5% and is transesterified using 8 to 42% w/w, of alcohol of general formula R-OH, where R represents (C_nH_{2n+1}) , n being any integer between 1 and 5, by known method in presence of not more than 0.5% w/w, of a known catalyst, at a temperature higher than the boiling point of the alcohol but not exceeding 215°C for not

Gas-Liquid Process, Thermodynamic Characteristics (19 Blends), Efficiency & Environmental Impacts,
SEM Particulate Matter Analysis and On-Road Bus Trial of a Proven NO$_x$ Less Biodiesel

5

less than 30 minutes under continuous turbulent condition at rpm in the range of 100-150 to get a mixture of ester and glycerol. The Reynolds number (N$_{Re}$) is maintained at not less than 4000 irrespective of the type of the reactor. The mixture of ester and glycerol is subjected to separation by known method for a period of not less than 4 hrs and the top layer ester is purified by conventional method for a period of not less than 8hrs. The process of separation as well as purification is repeated for not less than three times in succession to get biodiesel.

Fig. 1. Lab scale experimental setup

In lab scale experimental setup Fig.1, RBO was taken in the continuous stirred tank glass reactor (1 l) with reflex condenser, temperature control and agitation control setup. In another reactor, NaOH (50 g) was dissolved in methanol (300 ml). This solution was added slowly at the reactor maintained at 65-70 °C for 150 min. Then the entire mixture kept in the separating funnel. The top layer, biodiesel, is taken for the removal of methanol in the ROTO vacuum distiller. Then the methyl ester washed of distilled water (1 l) in the same reactor for 30 min. After washing, top layer in the separating funnel has to be washed with saline water for two times. Finally, clear biodiesel was kept in the oven for 4 h at 100°C. The ready to use biodiesel few samples shown in Fig.2.

Fig. 2. Ready to use biodiesel samples

In the bench scale level, Rice Bran Oil (RBO) experiments were carried out with standardized process conditions in high-pressure Parr Reactor (Fig.3.) inbuilt sophisticated controlling systems of reactor (20 l). Rice Bran Oil Biodiesel RBOBD (>150 l) was produced.

Fig. 3. Bench Scale lab Parr Reactor

In each lot, biodiesel sample has been analyzed for the conversion, fuel properties and composition. Quality consistency conformed by C_{13} and Proton of JEOL ECA 500 MHz NMR analysis and the composition by GCMS. All chemicals used were of LR/AR grade. A typical NMR spectrum show in Fig.4.

Gas-Liquid Process, Thermodynamic Characteristics (19 Blends), Efficiency & Environmental Impacts,
SEM Particulate Matter Analysis and On-Road Bus Trial of a Proven NO$_x$ Less Biodiesel

7

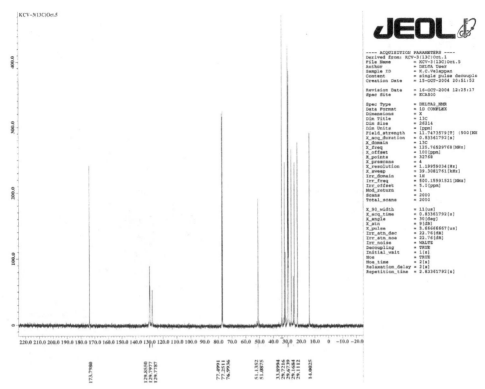

Fig. 4. A Typical C13, NMR Spectrum

The brief process description has been followed at the Pilot-scale preparation of biodiesel (Fig. 5. (a)), which was used for on-road trails from rice bran oil is following.

Rice bran oil is filtered to remove any impurities. 69 lit. of moisture free refined oil is taken in a Pilot Plant scale reactor (Fig. 5. (b)) of capacity 120 lit. Fitted with a reflux condenser and heated with agitation to 65°C. Then 345 gms of sodium hydroxide, 20.7 lit. of methanol are mixed separately and the mixture is slowly added to oil at 65°C.

The reaction mixture is mixed well, temperature is maintained at 65-70°C throughout the reaction and the reaction time is 150 min. When the reaction is complete, the contents are allowed to cool and transferred to a separating tank. After overnight settling, the mixture gets separated into two layers due to density difference.

The bottom layer-Glycerol is separated. The top layer - biodiesel is distilled at 65°C to recover unreacted alcohol. Then the methyl ester is washed for 30 minutes at 50°C with equal volumes of 0.1% dil. acetic acid to remove any traces of un reacted alkali. In case of emulsion formation after washing, saline water is used for second washing. The pH of the ester layer is adjusted to neutral while washing. After washing, the layers are allowed to settle for 30 min. The top layer is separated and biodiesel is dried in a pan drier for 2 hrs at 110°C. Then it is filtered to separate any traces of impurities. The final ready to use biodiesel product is found to be 60 lit.

Fig. 5. (a) Pilot-scale preparation of biodiesel (Fig. 5. (b))

Few thousand liters of Biodiesel produced in the pilot level which is used as feul in the on-Road bus trails. More than 26000 km exprimental trials were carried out in the Metropolitan Transport Corporation (MTC) buses in Chennai, Government of Tamil Nadu. Few clipings of MTC bus trails are shown in Fig.6. Initialy four buses have been taken for on-road trials in a single root but fuelled with different biodiesl percentage namely, B5, B10,B20 and B50. Then all the buses fuelled with 100% Biodiesel. The MTC, government of Tamil Nadu, has submitted the officeal report about the on –raod trials. The Fig. 7 showing the highligts signed by the MTC highre officails of the report in the reginal language namely TAMIL and Fig 8. Showing its translation in English.

3. Engine testing and exhaust gas analysis

RBOBD was tested in Kirloskar four stroke, single cylinder, water cooled, direct injection IC engine (Fig.9) with following parameters: bore, 80 mm; stroke, 110 mm; swept volume, 553 cm^3; clearance volume, 36.87 cm^3; compression ratio, 16.5:1; rated output, 3.7 kW at 1500 rpm; rated speed, 1500 rpm; injection pressure, 240 bar; fuel injection timing, 24 BTDC; type of combustion chamber, hemispherical open; lubricating oil, SAE 40; connecting rod length, 235 mm; valve diam, 33.7 mm; and maximum valve lift, 10.2 mm.

Gas-Liquid Process, Thermodynamic Characteristics (19 Blends), Efficiency & Environmental Impacts,
SEM Particulate Matter Analysis and On-Road Bus Trial of a Proven NO$_x$ Less Biodiesel

9

Fig. 5. (b) Pilot Plant scale reactor

Fig. 6. Few clipings of MTC bus trials

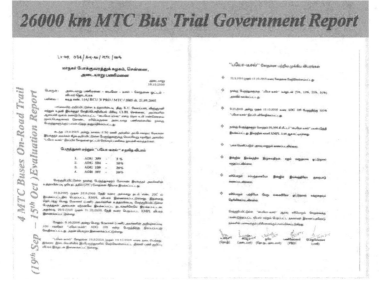

Fig. 7. Showing the highligts signed by the MTC highre officails

Gas-Liquid Process, Thermodynamic Characteristics (19 Blends), Efficiency & Environmental Impacts, SEM Particulate Matter Analysis and On-Road Bus Trial of a Proven NO$_x$ Less Biodiesel

11

Fig. 8. Highlights Translation in English

Fig. 9. A Test engine

DYNALOG, PCI 1050 system has been used for digital data acquisition during the engine trial. Online engine calibration (Fig.10) with a special software namey "Engine-soft". The very brief specifications are Number of channels (16); Resolution (12- bit A/D); Input range (± 10 V, ± 5V, 0 -10 V); Accuracy (0.025%) and Conversion time (8 μs).

Engine was coupled to a swinging field separating exciting type DC generator and loaded by electrical resistance bank to apply various load. An iron-constantan thermocouple measured exhaust gas temperature and mercury thermometer measured cooling water temperature. Carbon monoxide (CO), nitrous oxide (NO_x) and hydrocarbons (HC) were measured by DELTA 1600-L and MRU OPTRANS 1600, a fully microprocessor controlled system employing nondestructive IR technique. A U-tube manometer measured specific fuel consumption. TI diesel tune, 114-smoke density tester measured smoke particulate number.

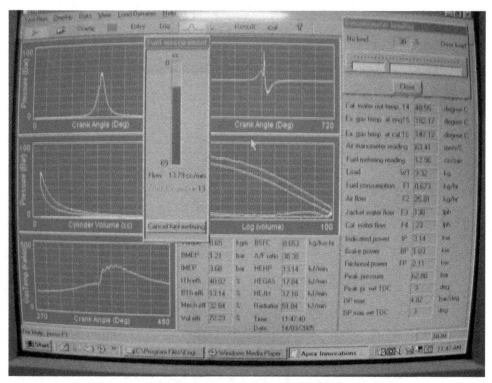

Fig. 10. Engine Calibration with "Engine-Soft"

The engine was started on neat diesel fuel and warmed up till liquid cooling water temperature was stabilized. During the performance of each trail, data were collected on time taken for 10 ml of fuel, load, exhaust gas temperature, cooling water inlet and outlet temperature, CO, CO_2, O_2, HC, NOx, smoke and sound. Graphical comparisons are described in the results and discussion. Smoke samples were collected in a white filter paper; this was taken for Scanning Electron Microscope (SEM) analysis to find the size of the particulate matter and to visualize the quantity of agglomeration. The SEM image is shown in Fig 10. Based on the data, specific fuel consumption, indicative thermal efficiency, brake thermal efficiency, mechanical efficiency and total fuel consumption were estimated. Similar procedures were repeated for RBOBD.

4. Results and discussion

4.1 Process conditions and compositions

RBOBD contains (GC-MS) esters of following acids: palmitic, 16; stearic, 2; oleic, 42; linoleic, 38; linolenic, 1.4; and arachidic, 0.6%. Quality consistency was conformed by C$_{13}$ and Proton of JEOL ECA 500 MHz NMR. Physico-chemical characteristics of RBOBD and its 19 blends (Table 3) show that most of the parameters comply with international standards of biodiesel. An NMR spectrum is already shown in Fig 4.

4.2 Comparison of Brake Power and Specific Fuel Consumption (SFC)

SFC of diesel, RBOBD and its various blends at different load (0-3.78 kW) were estimated and graphical representaion is shown in Fig 11. In comparison to diesel, a slight increase (10-15%) of SFC was found for RBOBD, B40, B50, B60 and B80 throughout all loads . At the maximum load (3.78 kW), SFC of B60 was found higher in comparison to the other blends. In particular to B20, the result shows that SFC was lower than diesel and other RBOBD and its blends in all the loads. The maximum increase (11.6%) was found at load 1.89 kW.

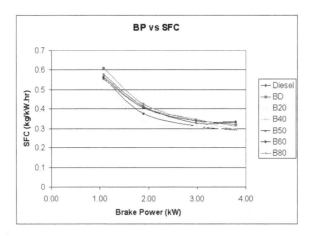

Fig. 11. Comparison of brake power and specific fuel consumption

4.3 Comparison of Brake Power and Fuel Consumption Time (FCT)

FCT of RBOBD and its various blends have been found less than the FCT of diesel, graphical representaion is shown in Fig 12. Slight decrease (5-10 %) of FCT was found for all fuels. Maximum decrease of FCT (12.5 %) was found at the brake power of 3.78 kW for B50 and B60. But, in particular, for B20, there was slight increase of FCT for the entire range of brake power. Maximum increase of FCT (12 %) was at 1.89 kW and minimum (3 %) at 3.78 kW.

Parameters	BD5	BD10	BD15	BD20	BD25	BD30	BD35	BD40	BD45	BD50
Acid value	0.27	0.3	0.31	0.49	0.5	0.57	3.49	0.74	0.54	0.87
Ash Content	0.0005	0.0006	0.0008	0.0010	0.0013	0.0051	0.0068	0.0051	0.0034	0.0038
Calcium	Nil	Nil	Nil	Nil	Nil	Nil	Nil	Nil	Nil	Nil
Carbon	82.27	82.17	82.07	81.97	81.90	81.85	81.80	81.90	81.98	82.01
Carbon residue (%)	0.00	0.0026	0.0030	0.0036	0.0041	0.0071	0.0096	0.011	0.01	0.012
Cetane Number	49	50	49	49	48	48	49	48	50	49
Cloud Point (C)	22	20	20	14	16	15	21	23	24	19
Density @ 15 C	0.8288	0.8335	0.8351	0.8391	0.8414	0.8437	0.8475	0.8520	0.8556	0.8588
Distillation 85	330	333	334	338	342	343	345	343	343	348
Distillation 95	344	347	347	350	356	358	360	358	360	361
Ester content	11.6	20.5	29.3	39.3	49.87	62.6	73.4	85.7	96.2	106.1
Flash Point	40	40	42	42	42	42	44	44	46	44
Free Glycerol	0.008	0.0091	0.010	0.011	0.009	0.012	0.014	0.013	0.016	0.015
G.C.V	10850	10720	10600	10610	10530	10500	10390	10300	10270	10210
Hydrogen	12.62	12.54	12.60	12.58	12.63	12.60	12.59	12.62	12.65	12.56
Iodine Value	10.2	14.1	17.9	23.3	28.7	32.5	35.5	42.3	46.2	49.5
N.C.V.	10181	10055	9932	9943	9860	9832	9723	9631	9600	9544
Nitrogen	0.031	0.030	0.030	0.034	0.031	0.033	0.034	0.032	0.030	0.032
Oxygen	5.063	5.245	5.285	5.402	5.426	5.505	5.564	5.436	5.33	5.308
Phosphorus	0.0006	0.00071	0.010	0.0015	0.0023	0.0029	0.0034	0.0043	0.0051	0.0058
Potassium	2.0	1.9	1.8	1.7	1.8	1.65	1.70	1.90	1.80	1.6
Pour point C	-22	-20	-20	-18	-15	-15	-16	15	-14	-13
Sodium	2.1	2.0	1.8	1.9	1.6	1.8	1.50	1.50	1.60	1.4
Sulphated Ash	0.0010	0.0013	0.0015	0.0022	0.0031	0.0068	0.0094	0.0064	0.0050	0.0064
Sulphur	0.016	0.015	0.015	0.014	0.013	0.012	0.012	0.012	0.010	0.009
Sulphur	0.016	0.015	0.015	0.014	0.013	0.012	0.012	0.012	0.010	0.009
Total contamination	0.096	0.009	0.011	0.011	0.011	0.012	0.009	0.0098	0.011	0.013
Total Glycerol	0.013	0.018	0.019	0.022	0.025	0.033	0.036	0.043	0.051	0.054
Viscosity @ 40 C	2.6	2.7	2.9	3.0	3.1	3.4	3.4	3.5	3.7	3.9
Water & sediments	0.022	0.026	0.048	0.029	0.024	0.027	0.030	0.029	0.029	0.031
Water content	0.021	0.024	0.045	0.026	0.022	0.022	0.026	0.027	0.027	0.0281
Acid value	1.08	0.97	1.07	1.08	1.2	1.27	1.4	1.41	1.43	1.54

Parameters	BD5	BD10	BD15	BD20	BD25	BD30	BD35	BD40	BD45	BD50
Ash Content	0.0037	0.0047	0.0045	0.0039	0.0076	0.0080	0.0081	0.0086	0.0093	0.0096
Calcium	Nil	Nil	Nil	Nil	Nil	Nil	-	-	-	-
Carbon	83.17	85.10	83.01	82.96	82.90	82.79	82.80	82.70	82.56	82.42
Carbon residue (%)	0.02	0.021	0.031	0.019	0.021	0.016	0.025	0.029	0.029	0.036
Cetane Number	49	50	48	49	48	47	47	46	46	45
Cloud Point Deg.C	9	9	12	14	14	13	16	14	19	23
Density @ 15 Deg.C	0.8625	0.8635	0.8666	0.8704	0.8748	0.8794	0.8814	0.8854	0.8889	0.8901
Distillation 85	342	326	340	342	344	345	344	343	347	348
Distillation 95	350	342	350	352	356	357	356	355	359	361
Ester content	111.7	120.2	130.3	132.2	150.5	156.2	164.9	164.9	181.1	192
Flash Point	43	46	64	60	64	70	68	90	90	124
Free Glycerol	0.016	0.015	0.018	0.017	0.019	0.018	0.020	0.021	0.024	0.025
G.C.V	10190	10090	10050	9970	9920	9860	9750	9690	9610	9810
Hydrogen	12.84	12.94	12.89	12.80	12.76	12.80	12.70	12.69	12.72	12.60
Iodine Value	51.3	56.2	57.2	63.2	67.7	72.7	77.3	80.7	85.6	86.4
N.C.V.	9509	9364	9367	9242	9284	9241	9187	9077	9016	8942
Nitrogen	0.030	0.032	0.031	0.029	0.028	0.028	0.029	0.029	0.031	0.030
Oxygen	3.936	3.904	3.904	4.188	4.271	4.362	4.451	4.563	4.672	4.933
Phosphorus	0.062	0.064	0.0068	0.0078	0.0076	0.0084	0.0083	0.0074	0.0095	0.0097
Potassium	1.2	2.0	2.1	2.1	2.0	2.2	2.9	2.2	2.1	1.9
Pour point Deg.C	-13	-13	-12	-12	-11	-11	-10	-10	-9	-8
Sodium	3.1	3.6	3.6	3.5	3.4	3.7	3.4	3.0	2.9	2.2
Sulphated Ash	0.0073	0.0080	0.0064	0.0057	0.0081	0.0061	0.0069	0.0053	0.012	0.013
Sulphur	0.024	0.024	0.023	0.023	0.021	0.020	0.020	0.018	0.017	0.017
Sulphur	0.024	0.024	0.023	0.023	0.021	0.020	0.020	0.018	0.017	0.017
Total contamination	0.015	0.018	0.022	0.021	0.024	0.025	0.029	0.030	0.035	0.044
Total Glycerol	0.064	0.069	0.070	0.074	0.075	0.083	0.088	0.096	0.099	0.10
Viscosity @ 40 Deg.C	4.0	4.2	4.40	4.60	4.8	5.0	5.2	5.5	5.7	6.0
Water & sediments	0.014	0.018	0.021	0.021	0.0034	0.024	0.028	0.031	0.035	0.041
Water content	0.012	0.017	0.020	0.020	0.021	0.022	0.028	0.029	0.035	0.041

Table 3. Physico-chemical characteristics of RBOBD and its 19 blends

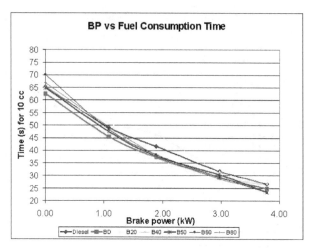

Fig. 12. Comparison of brake power and fuel consumption time

4.4 Comparison of Brake Power and Total Fuel Consumption (TFC)

TFC increased with increase of brake power, graphical representaion is shown in Fig 13. A maximum increase (13%) was at the load 1.89 kW. TFC's of RBOBD blends, B40, B50, B60 and B80, are higher (5-10%) than the TFC of diesel. But B20's TFC is slightly lesser than diesel and all the other RBOBD blends from the minimum load to the maximum load. Maximum TFC decrease (10%) was observed for B20 at 1.89 kW. Overall trend shows that the percentage decrease in TFC is inversely proportional to the brake power. At the maximum load, the increasing order of TFC is B20, Diesel, B40, B80, RBOBD and B60.

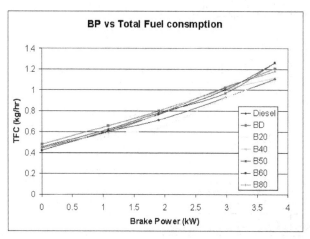

Fig. 13. Comparison of brake power and total fuel consumption

4.5 Comparison of Brake Power and Exhaust Gas Temperature (EGT)

EGT increases with increase of brake power, graphical representaion is shown in Fig 14.

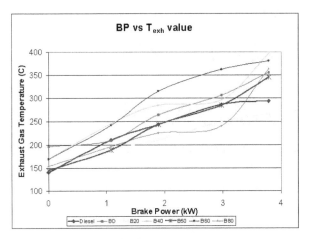

Fig. 14. Comparison of brake power and exhaust gas Temperatur

In comparison with EGT of diesel in each load, EGT of RBOBD and all the blends were higher. The highest value of EGT (395°C) was found with B40 at the maximum load of 3.78 kW, whereas corresponding value of normal diesel was 294°C only. Percentage increase of EGT of RBOBD decreased with the increase of load. Maximum increase (40%) of EGT was found at lower load of zero brake power. EGT of B40, B50, B60 and B80 were 350-400°C at the maximum load. The percentage increase (20-40%) of EGT of B60 was higher than all the loads. EGT of B20 was found to be slightly lower than EGT of normal diesel in all loads (0-3.78 kW). The minimum EGT decrease (9.5%) and maximum decrease (16.3%) of RBOBD and blends were found at 1.1 Kw and 2.98 kW respectively as compared to diesel. EGT of B20, B50 and B80 were found to be lower than EGT of diesel at 0-2.97 kW.

4.6 Comparison of Brake Power and Brake Thermal Efficiency (BTE)
BTE increases with increase of load, graphical representaion is shown in Fig 15.

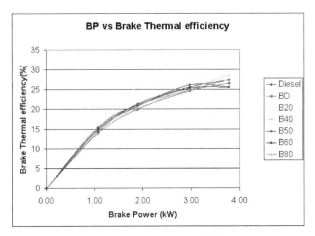

Fig. 15. Comparison of brake power and brake thermal efficiency

BTE of RBOBD was less (5%) than diesel with respect to all loads. All other RBOBD blends (B40, B50, B60 and B80) were within 5 % only. Maximum reduction (20%) of BTE was found at 1.89 kW and minimum increase (7%) at 2.9765 kW. BTE of B20 was found higher than BTE of normal diesel in all loads. At maximum load (3.78 kW), BTE for B20 (29.7%), B40 (28.6%), B50 (25.6%) and B60 & B80 (25.5%) are higher than BTE of diesel.

4.7 Comparison of Brake Power and Indicative Thermal Efficiency (ITE)

ITE of RBOBD, B50, B60 and B80 were found lower than ITE of diesel, graphical representaion is shown in Fig 16 . ITE of B20 and B40 were slightly more than ITE of diesel. Maximum ITE (57.9%) was found for B20 at 1.89 kW. Reduction (5-15%) was observed in ITE of various blends. But the ITE of all fuels shows that values are almost steady throughout the entire brake power.

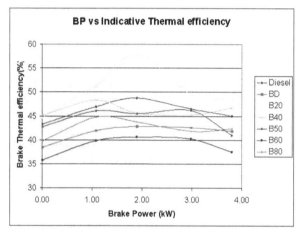

Fig. 16. Comparison of brake power and indicative thermal efficiency

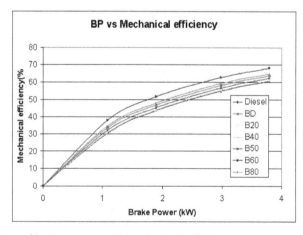

Fig. 17. Comparison of brake power and mechanical efficiency

4.8 Comparison of Brake Power And Mechanical Efficiency (ME)

ME of the engine run with RBOBD and various blends increases with increase of brake power in comparison with normal diesel, graphical representaion is shown in Fig 17. ME of RBOBD was less (5%) than diesel with respect to all loads. There was a slight increase of ME for all RBOBD blends in the order of B40, B50, B20, B80 and B60. Highest ME (68%) observed for B60 was at the maximum load. Overall trend shows that percentage increase of ME was decreased with increase of load. The result of B60 shows that the minimum increase (12%) was found at maximum load and maximum increase (24%) at minimum load.

4.9 Comparison of Brake Power and Hydrocarbons (HC)

HC increased with increase of brake power, graphical representaion is shown in Fig 18. RBOBD and its five blends showed lower HC (50-100%) than diesel. Reduction of HC (60%) of RBOBD and its blends are at the maximum load; B50 shows higher HC reduction than other blends at all loads.

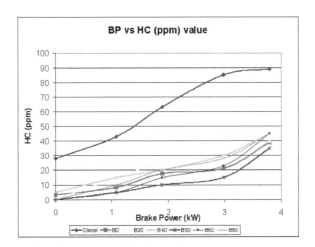

Fig. 18. Comparison of brake power and hydrocarbons

4.10 Comparison of Brake Power and CO emission

CO emission increased with increase of brake power, graphical representaion is shown in Fig 19. There was decrease of CO emission (50-80%) of RBOBD and its all five blends in comparison with CO emission of diesel. Reduction (>50%) of CO emission was found at 2.97 kW. Within blends, B20 shows lower CO emission (70-80%), which decreased with increase of load.

4.11 Comparison of Brake Power and CO$_2$ emission

CO$_2$ emission increased with increase of brake bower, graphical representaion is shown in Fig 20. There was slight increase in CO$_2$ emission of RBOBD and its blends as compared to diesel. More variation of percentage increase was found within all RBOBD blends at the load 1.89 kW. The overall trend shows that the CO$_2$ emissions are similar to diesel at each load.

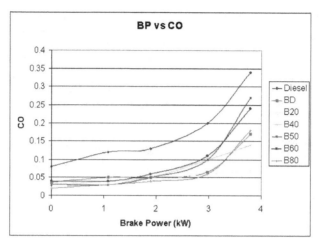

Fig. 19. Comparison of brake power and carbon monoxide emission

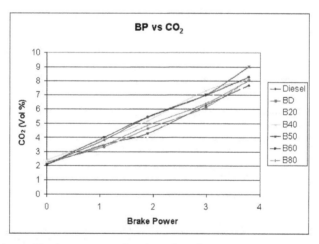

Fig. 20. Comparison of brake power and carbon dioxide emission

4.12 Comparison of Brake Power and NO_x emission

NO_x increased with increase of brake power, graphical representaion is shown in Fig 21. There was a reduction (10-55%) of NO_x of RBOBD and its blends in comparison with NO_x values of diesel in each load. The trend shows that at minimum load, percentage reduction was maximum and at the maximum load, the percentage reduction of NO_x was minimum. The percentage reduction of NO_x decreased with increase of brake power. NO_x values at maximum load (3.78 kW) were found to be: diesel, 942; B80, 858; B50, 782; RBOBD, 753; B20, 677; and B60, 660 ppm. At 3.78 kW, maximum reduction (28%) was found for B20 and minimum (8.9%) for B80.

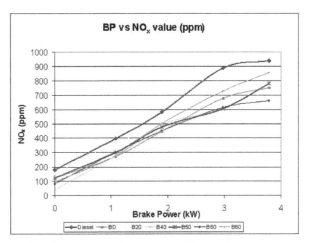

Fig. 21. Comparison of brake power and nitrogen oxides emission

4.13 Comparison of Brake Power and O_2

O_2 decreased with increase of brake bower, graphical representaion is shown in Fig 22. Deviations (5-10%) were found for RBOBD and its blends. At maximum load, O_2 (6.5) in B50 was less (25%) than O_2 (9.1) of diesel.

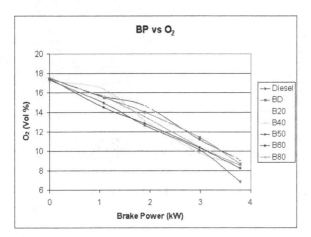

Fig. 22. Comparison of brake power and Oxygen

4.14 Comparison of Brake Power and sound

Sound or noise increased with increase of load, graphical representaion is shown in Fig 23. Sound values of RBOBD and its blends are found lower (15-30%) than the sound values of diesel throughout the brake power. Within comparison of RBOBD and its blends, there was not much change in sound in all the loads. The minimum decrease (13.6%) was observed at the minimum load, and the maximum decrease (30%) at the maximum load (3.78 kW). At the higher load, sound reduction (21-30%) for RBOBD and all of its blends compared to diesel.

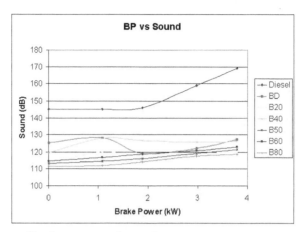

Fig. 23. Comparison of brake power and sound

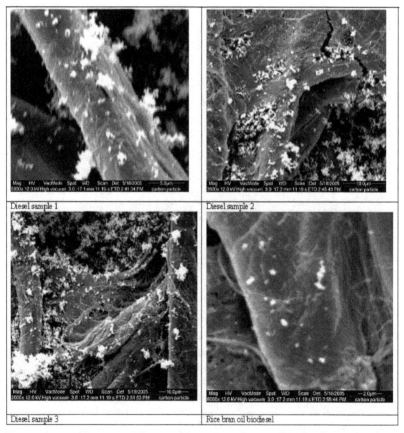

Fig. 24. Comparison of Scanning Electron Microscope Image (SEM) of Diesel and RBOBD

4.15 Comparison of SEM image

During the engine trails, smoke was collected in the white filter paper. Few samples at the maximum load were taken for SEM analysis at different resolutions. SEM shows much reduction in the particulate matter in the biodiesel as compared to diesel (Fig. 24). The particle size in the smoke of RBOBD is less than 0.5 μm.

5. Conclusions

This is most ambitious and successful technology development initiative for alternative energy options, which is the important global agenda, and will be good for the environment. These are the key component for energy security and have positive economic, social and environmental impacts. The conventional fossil fuel energy sources are the major cause of climatic changes, this biodiesel leads to minimize the emission to the environment and sustainable society. The biodiesel production with this technology may cut fossil fuel imports and dependency and thus, free up funds that can be invested in social and economic development. This is the process innovation for the less NO$_x$ emitting Biodiesel. An international Patent has been filed through the Intellectual Property Management Division of Council of Scientific and Industrial Research (CSIR), New Delhi in the year 2003. *(Ref No. IMPD. 0290 NF 2003, PCT / IB 03/ 05349 (21.11.2003, US & JP 20050108927 and World Intellectual Property Organization-WO/2005/052103)., Patent granted Firest in Singapore, Patent Number 119411 on 31st May 2006 and then Australia , patent Number-2003282270 and Sri Lanka Patent Number - 13950.* In India, this is the first technological breakthrough in On-Road trials fuelled with biodiesel were successfully carried out in state government metropolitan Transport corporation (MTC) buses more than 26000 km without any engine modifications. There are 15% increases in the KMPL found after the trials. The MTC bus trials prove that this Biodiesel is substitute for diesel in the real conditions. On-Road trials fuelled with 100% Biodiesel (free biodiesel samples) by owner-driven in different vehicle models, namely, *Toyota Qualis, Mitsubishi Lancer, Bolero, Hyundai Acent etc.*

RBOBD was obtained by the optimized process conditions of transesterification and purification. RBOBD (> 150 l) is produced in Bench-Scale Parr reactor. RBOBD showed 15% increase in SFC, 25% increase in BTE, less exhaust gas temperature, 12% increase in FCT, more ITE and ME, 60% reduction in HC, more than 75% reduction in CO, minimum of 10% reduction in NO$_x$ and 30 % sound reduction. The combustion of RBOBD and its blends are found to be lesser pollutants than compared to diesel. SEM shows reduction of particulate matter and the size of the solid particle is less than 0.5 μm. This study proves that RBOBD is environment friendly alternate fuel for diesel without any engine modification.

6. Acknowledgements

Author acknowledges Director, Central Leather Research Institute, Chennai, for all support and Director, Indian Institute of Petroleum, Dehradun for the support provided for all the fuel properties analysis. The efforts of Dr Nagarajan, Anna University, Chennai for engine trails are highly acknowledged. The author acknowledges The Managing Director and other officials of MTC, Government of Tamil Nadu for their support provided for the successful on-road trials in buses.

7. References

Agarwal A K, Vegetable oils verses diesel fuels: Development and use of biodiesel in compression ignition engine, *TIDE*, 8 (1998) 191-204.

Antolin G, Tinaut F V, Briceno Y, Castano V, Perez C & Ramirez A I, Optimization of biodiesel production by sunflower oil transesterification, *Biores Technol*, 83 (2002) 111-114.

Arumugam C, Kumar S, Rajam L & Sundarresan A, Integrated technology for refining rice bran oil and by-products recovery, *6thNat Semin on Rice Bran Oil*, Lucknow, 2003.

Barnwal B K & Sharma M P, Prospects of biodiesel production from vegetable oils in India, *Renewable & Sustainable Energy Rev*, 9 (2005) 363-378.

De-Gang Li, Huang Zhen, Lu Xingcai, Zhang Wu-gao & Yang Jian-guang, Physico-chemical properties of ethanol-diesel blend fuel and its effect on performance and emissions of diesel engines, *Renewable Energy*, 30 (2005) 967-976.

Fernando Neto da Silva, Antonio S P & Jorge R T, Technical feasibility assessment of oleic sunflower methyl ester Utilisation in diesel bus engine, *Energy Convers & Manage*, 44 (2003) 2857-2878.

Freedam B, Butterfield R O & Pryde E H, Transesterification kinetics of soyabean oil, *J Am Oil Chem Soc*, 63 (1986) 1375-1380.

Goering C E, Schwab A, Doughherty M, Pryde M & Heakin A, Fuel properties of eleven vegetable oils, *Trans ASAE*, 85 (1982)1472-1483.

Michael S G & Robert L M, Combustion of fat and vegetable oil derived fuels in diesel engines, *Prog Energy Combust Sci*, 24(1998) 125-164.

Mohamad I, Al-Widyan & Ali O A, Experimental evaluation of the transesterification of waste palm oil into biodiesel, *Biores Technol*, 85 (2002) 253-256.

Mohamad I, Al-widyan, Ghassan T & Moh'd A Q, Utilization of ethyl ester of waste oils as fuel in diesel engines, *Fuel Processing Technol*, 76 (2002) 91-103.

Noureddini H & Zhu D, Kinetics of transesterification of soybean oil, *J Am Chem Soc*, 74 (1997) 1457-1463.

Prasad Rao G A & Rama Mohan P, Effect of supercharging on performance of a DI diesel engine with cotton seed oil, *Int J Energy Conserv & Manage*, 44 937-940.

Roger A K & Jaiduk J O, A rapid engine test to measure injector fouling in diesel engine using vegetable oil fuels, *J Am Oil Chem Soc*, 62 (1985) 1563-1564.

Sinha S & Misra N C, Diesel fuel alternative from vegetable oils, *Chem Engg World*, 32 (1997) 77-80.

Srivastava A & Prasad R, Triglycerides-based diesel fuels, *Renewable & Sustainable Energy Rev*, 4 (2000) 111-133.

Sukumar P, Vedaraman N, Rambrahamam B V & Nagarajan G, Mahua (*Madhuca indica*) seed oil: A source of renewable energy in India, *J Sci Ind Res*, 64 (2005) 890-896.

Yi-Hsu J & Shaik R V, Rice bran oil as potential resource for biodiesel: A review, *J Sci Ind Res*, 64 (2005) 866-882.

2

Alternative Methods for Fatty Acid Alkyl-Esters Production: Microwaves, Radio-Frequency and Ultrasound

Paula Mazo, Gloria Restrepo and Luis Rios

Universidad de Antioquia. Grupo Procesos Fisicoquímicos Aplicados
Colombia

1. Introduction

Biodiesel production is a very modern and technological area for researchers due to the relevance that it is winning every day because of the increase in the petroleum price and the environmental advantages (Mustafa, 2011).

Biodiesel is a mixture of mono-alkyl esters of long chain fatty acids, is an alternative fuel made from renewable sources as vegetable oils and animal fats. It is biodegradable, non-toxic, show low emission profiles and also is beneficial environmentally. (Fangrui and Milford, 1999).

Biodiesel is quite similar to petroleum-derived diesel in its main characteristics such as cetane number, energy content, viscosity, and phase changes. Biodiesel contains no petroleum products, but it is compatible with conventional diesel and can be blended in any proportion with fossil-based diesel to create a stable biodiesel blend. Therefore, biodiesel has become one of the most common biofuels in the world (Lin et al., 2011). There are four primary techniques for biodiesel production: direct use and blending of raw oils, micro-emulsions, thermal cracking and trans-esterification (Siddiquee and Rohani, 2011).

Direct use of vegetable oil and animal fats as combustible fuel is not suitable due to their high kinematic viscosity and low volatility. Furthermore, its long term use posed serious problems such as deposition, ring sticking and injector chocking in engine. Microemulsions with alcohols have been prepared to overcome the problem of high viscosity of vegetable oils. Another alternative way to produce biodiesel is through thermal cracking or pyrolysis. However, this process is rather complicated to operate and produce side products that have not commercial value. The most commonly used method for biodiesel production is trans-esterification (also known as alcoholysis) reaction in presence of a catalyst. Trans-esterification is the process of exchanging the alkoxy group of an ester compound with another alcohol (Lam et al., 2010).

Esterification is the sub category of trans-esterification. This requires two reactants, carboxylic acids (fatty acids) and alcohols. Esterification reactions are acid-catalyzed and proceed slowly in the absence of strong acids such as sulfuric, phosphoric, sulfonic-organic acids and hydrochloric acid (Vyas et al., 2010).

The fatty acid methyl esters (FAME) are more used because of its facility of production, however, presents operating problems at low temperatures for its high content of saturated fractions that crystallize and can block the filters of the engines. One of the alternatives to reduce the flow properties at low temperatures (FPLT) of methyl esters specially the obtained from oil palm is use alkyl esters, obtained through of trans-esterification with branched alcohols, that prevent the agglomeration and formation of crystals of these methyl esters.

Alkyl esters can be produced through trans-esterification of triglycerides, which are separated by immiscibility and higher density. (Marchetti et al., 2007; Ma and Hanna, 1999; Vicente et al., 2004)

Very few studies have been made with the aim to obtain alkyl esters and all are obtained by homogeneous catalysis (Lee et al., 1995). Yields of these reactions are very low by the high steric hindering that presenting the branched alcohols. To increase the conversion, in this work, we propose use assisted reactions by alternative methods.

The preparation of fatty acid alkylester using alternative methods, such as: electromagnetic radiation (microwave, radio frequency) and ultrasound, offers a fast, easy route to this valuable biofuel with advantages of a short reaction time, a low reactive ratio, an ease of operation a drastic reduction in the quantity of by-products, and all with reduced energy consumption.

In this work the revision of the relevant aspects of the production optimization, intrinsic effects and parameters more relevant in the synthesis and characterization of fatty acid alkylesters (biodiesel) using as alternative methods: Microwaves, Radio Frequency and Ultrasound is proposed.

2. Fatty acid alkylesters production assisted by microwaves

Electromagnetic radiation (EMR) is a form of energy exhibiting wave like behaviour as it travels through space. EMR has both electric and magnetic field components, which oscillate in phase perpendicular to each other and perpendicular to the direction of energy propagation. Electromagnetic radiation is classified according to the frequency of its wave. In order of increasing frequency and decreasing wavelength, these are radio waves, microwaves, infrared radiation, visible light, ultraviolet radiation, X-rays and gamma rays (Serway and Jewett, 2004).

Microwaves belong to the portion of the electromagnetic spectrum with wavelengths from 1 mm to 1 m with corresponding frequencies between 300 MHz and 300 GHz.

Within this portion of the electromagnetic spectrum there are frequencies that are used for cellular phones, radar, and television satellite communications. For microwave heating, two frequencies, reserved by the Federal Communications Commission (FCC) for industrial, scientific, and medical (ISM) purposes are commonly used for microwave heating. The two most commonly used frequencies are 0.915 and 2.45 GHz. Recently, microwave furnaces that allow processing at variable frequencies from 0.9 to 18 GHz have been developed for material processing (Thostenson and Chou, 1999). Microwave radiation was discovered as a heating method in 1946, with the first commercial domestic microwaves being introduced in the 1950s. The first commercial microwave for laboratory utilization was recognized in 1978 (Gedye et al., 1986; Giguere et al., 1986).

Over the last decade, microwave dielectric heating as an environmentally benign process has developed into a highly valuable technique, offering an efficient alternative energy source for numerous chemical reactions and processes. It has many advantages compared to conventional oil-bath heating, such as non-contact heating, energy transfer instead of heat transfer, higher heating rate, rapid start-up and stopping of the heating, uniform heating with minimal thermal gradients, selective heating properties, reverse thermal effects (heating starting from the interior of the material body), energy savings and higher yields in shorter reaction time (Tierney and Lidstrom, 2005). Microwave heating is based dielectric heating, the ability of some polar liquids and solids to absorb and convert microwave energy into heat. In this context, a significant property is the mobility of the dipoles by either ionic conduction or dipolar polarization and the ability to orient them according to the direction of the electric field. The orientation of the dipoles changes with the magnitude and the direction of the electric field. Molecules that have a permanent dipole moment are able to align themselves through rotation, completely or at least partly, with the direction of the field. Therefore, energy is lost in the form of heat through molecular friction and dielectric loss (Loupy, 2002). The amount of heat produced by this process is directly related to the capability of the matrix to align itself with the frequency of the applied electric field. If the dipole does not have enough time to realign, or reorients too rapidly with the applied field, no heating occurs (Kappe, 2004).

The production of biodiesel via the conventional heating system appears to be inefficient due to the fact that the heat energy is transferred to the reactants through conduction, convection and radiation from the surface of the reactor. Hence, conventional heating requires longer reaction time and a larger amount of heat energy to obtain a satisfactory biodiesel. The replacement of conventional heating by microwave radiation for the transesterification process is expected to shorten the reaction time due to the transfer of heat directly to the reactants. The microwave radiation during the transesterification process is expected to create (i) an alignment of polar molecules such as alcohols with a continuously changing magnetic field generated by microwaves and (ii) molecular friction due to which heat will be generated (Yaakob et al., 2009).

The involvement of such heterogeneous catalytic systems under microwave conditions represents an innovative approach with processing advantages. These solid-state catalysts find scope in the context of green chemistry development as they are active in solvent free or dry media synthesis, with potential advantages in terms of separation, recovery post-reaction and recycling assays. The creation of hot spots, specific under MW conditions, is typically utilized for energy saving as improved yields and selectivities are recorded after shorten reaction times at lower nominal temperatures. These hot spots may induce a re-organization of the catalyst under microwave conditions and are probably responsible for reaction rates and selectivity enhancement (compared to conventional heating at the same nominal temperature) (Richela et al., 2011)

2.1 Esterification reactions assisted by MW

The esterification reaction is a slow equilibrium, and can be catalyzed by Brønsted acids such as sulfuric acid. The main problem is the generation of highly acidic waste causing a serious environmental problem, and to reduce this problem have been used alternative heterogeneous catalysts and microwaves as a heating source to promote and increase the yielding. Algunos catalizadores empleados son: scandium triflate and bismuth triflate (Socha and Sello, 2010), sulfated zirconia (Kim et al., 2011a), niobium oxide (Melo et al., 2010), entre otros.

The temperature presented a pronounced effect on the conversion, following an exponential dependence. The results for a distinct molar ratio of alcohol/fatty acid indicated that the increase of this parameter lead to a decrease on the reaction conversion. In general, the esterification reaction under microwave irradiation yielded similar results to those obtained with the conventional heating but with very fast heating rates (Melo et al., 2009). The pulsed microwaves with repetitive strong power could enhance the efficiency of biodiesel production relative to the use of continuous microwave with mild power (Kim et al., 2011b). Electric energy consumption for the microwave heating in this accelerated esterification was only 67% of estimated minimum heat energy demand because of significantly reduced reaction time (Kim et al., 2011a).

For oils with a high content of free fatty acid FFA as palm oil, has been proposed obtain alkyl ester from crude palm oil (CPO), using microwaves like heating source, in a process of two stages by means of homogeneous and heterogeneous catalysis; the first stage (esterification), was made using sulfuric acid and Dowex 50X2, Amberlyst 15 and Amberlite IR-120 resin catalysts, to diminish the acid value of the oil, avoiding the soap formation and facilitating the separation of the phases. In these works has been reported the obtaining of alkyl ester using alcohols non-conventional such as: ethanol (EtOH) (Suppalakpanya et al., 2010a, 2010b), isopropyl (IsoprOH), isobutyl (IsobuOH), 2-butyl (2-BuOH) and Isopentyl (IsopentOH) alcohols (Mazo and Rios, 2010a; Mazo and Rios, 2010b), where was found that that the acidity order obtained for the catalysts is Dowex < Amberlite < Amberlyst, and the order for the alcohols: Methanol < isopropyl alcohol < isobutyl alcohol < 2-butyl alcohol < isopentyl alcohol, because Dowex microreticular resin presents the lowest divinylbenzene (2%), which has a lower cross-linking that produces a high expansion of the resin in a polar medium, and the resin can expand their pores up to 400%, enabling the income of the voluminous substrate (FFA) and its protonation. Amberlyst 15 macroreticular resin is activated due to its surface area, and the protons located on the outer surface seem that catalyse the esterification because the interiors are inaccessible due to high cross-linking. The reaction is favoured with the increasing of polarity of solvents.

Table 1 shows the work carried out for bio-diesel production by esterification of FFA under different conditions using microwave irradiation.

2.2 Transesterification reactions assisted by MW

Vegetable oils are becoming a promising alternative to diesel fuel because they are renewable in nature and can be produced locally and in environmentally friendly ways. Edible vegetable oils such as canola and soybean oil in the USA, palm oil in Malaysia, rapeseed oil in Europe and corn oil have been used for biodiesel production and found to be good diesel substitutes. Non-edible vegetable oils, such as Pongamia pinnata (Karanja or Honge), Jatropha curcas (Jatropha or Ratanjyote), Madhuca iondica (Mahua) and Castor Oil have also been found to be suitable for biodiesel production (Yusuf et al., 2011).

Transesterification (also called alcoholysis) is the reaction of a fat or oil with an alcohol (with or without catalyst) to form esters and glycerol. Since the reaction is reversible, excess alcohol is used to shift the equilibrium to the product side (Fangrui and Milford, 1999). Under Transesterification reaction with alcohol the first step is the conversion of triglycerides to diglycerides, which is followed by the subsequent conversion of higher glycerides to lower glycerides and then to glycerol, yielding one methyl ester molecule from each glyceride at each step (Hideki et al., 2001).

FFA	Catalyst	Catalyst amount (%)	Alcohol	Oil to alcohol molar ratio	Microwave reaction conditions	Ester conversion (%)	Ref.
Oleic	-	-	MeOH EtOH	1:10	Synthos 3000-Anton Paar. 1400W. 30min, 200°C	51.8 31.5	(Melo et al., 2009)
Linoleic	-		MeOH			49.6	
Oleic	Sulfated Zirconia	5 wt	MeOH	1:20	Experimental MW heating system 20min, 60°C	90.0	(Kim et al., 2011a)
Oleic	Amberlyst 15 dry	10 wt	MeOH	1:20	Experimental MW heating system. Pulsed MW. 15min, 60°C	66.1	(Kim et al., 2011b)
Oleic	Niobium Oxide Sulfated Zirconia	5 wt	MeOH	1:10	Synthos 3000-Anton Paar. 1400W. 20min, 200°C	68.0 68.7	(Melo et al., 2010)
Linoleic Linoleic Oleic Oleic Myristic Myristic Palmitic Palmitic	$Sc(OT_f)_3$ $Bi(OT_f)_3$ $Sc(OT_f)_3$ $Bi(OT_f)_3$ $Sc(OT_f)_3$ $Bi(OT_f)_3$ $Sc(OT_f)_3$ $Bi(OT_f)_3$	1%mol	MeOH	48 eq	Biotage MW reactor. 1min, 150°C	97.0 98.0 100.0 88.0 98.0 90.0 100.0 99.0	(Socha and Sello, 2010)
FFA Palm Oil	H_2SO_4	2.5%wt Oil	MeOH IsoprOH IsoBuOH 2-BuOH IsopentOH	1:8	Domestic MW 1000W 60 min, 60°C 60 min, 75°C 60 min, 105°C 60 min, 90°C 60 min, 115°C	99.8 99.8 96.2 95.5 90.8	(Mazo and Rios, 2010a)
FFA Palm Oil	H_2SO_4	4% wt FFA	EtOH	1:24	Domestic MW 800W 60 min, 70°C	87.7	(Suppalakpanya et al., 2010a and b)
FFA Palm Oil	Dowex 50X2	10%wt Oil	MeOH IsoprOH IsoBuOH 2-BuOH IsopentOH	1:20	Domestic MW 1000W 60 min, 60°C 60 min, 75°C 60 min, 105°C 60 min, 90°C 60 min, 115°C	95.6 86.2 82.8 78.5 77.7	(Mazo and Rios, 2010b)
FFA Palm Oil	Amberlite IR120	10%wt Oil	MeOH IsoprOH IsoBuOH 2-BuOH IsopentOH	1:20	Domestic MW 1000W 60 min, 60°C 60 min, 75°C 60 min, 105°C 60 min, 90°C 60 min, 115°C	91.3 85.1 81.4 74.8 74.1	(Mazo and Rios, 2010b)
FFA Palm Oil	Amberlyst15	10%wt Oil	MeOH IsoprOH IsoBuOH 2-BuOH IsopentOH	1:20	Domestic MW 1000W 60 min, 60°C 60 min, 75°C 60 min, 105°C 60 min, 90°C 60 min, 115°C	91.4 84.7 80.6 66.8 73.5	(Mazo and Rios, 2010b)

Table 1. Microwave assisted esterification.

Several examples of microwave irradiated transesterification methods have been reported using homogenous alkali catalyst (Kumar et al., 2011; Azcan and Danisman, 2008), acid catalyst (Mazo and Rios, 2010a) and heterogeneous alkali catalyst (Patil et al., 2011), heterogeneous acid catalyst (Yuan et al., 2009) and enzymatic (Yu et al., 2010).

Microwave synthesis is not easily scalable from laboratory small-scale synthesis to industrial multi kilogram production. The most significant limitation of the scale up of this technology is the penetration depth of microwave radiation into the absorbing materials, which is only a few centimeters, depending on their dielectric properties. The safety aspect is another reason for rejecting microwave reactors in industry (Groisman and Aharon, 2008).

The preparation of biodiesel using a scientific microwave apparatus offers a fast, easy route to this valuable biofuel with advantages of a short reaction time, a low oil/methanol ratio, and an ease of operation. The methodology allows for the reaction to be run under atmospheric conditions; it is complete in a matter of a few minutes and can be performed on batch scales up to 3 kg of oil at a time (Leadbeater and Stencel, 2006).

The continuous-flow preparation of biodiesel using a commercially available scientific microwave apparatus offers a fast, easy route to this valuable biofuel. The methodology allows for the reaction to be run under atmospheric conditions and performed at flow rates of up to 7.2 L/min using a 4 L reaction vessel. Energy consumption calculations suggest that the continuous-flow microwave methodology for the transesterification reaction is more energy-efficient than using a conventional heated apparatus (Barnard et al., 2007).

Few studies report the use of alcohols different to methanol. Alcohols more used are ethanol and butanol, and the latter is a versatile and sustainable platform chemical that can be produced from a variety of waste biomass sources. The emergence of new technologies for the production of fuels and chemicals from butanol will allow it to be a significant component of a necessarily dynamic and multifaceted solution to the current global energy crisis. Recent work has shown that butanol is a potential gasoline replacement that can also be blended in significant quantities with conventional diesel fuel (Harvey and Meylemans, 2011).

Table 2 shows the work carried out for bio-diesel production from various feedstocks, catalysis and alcohols under different conditions using microwave irradiation.

Oil	Catalyst	Catalyst amount (%wt)	Alcohol	Oil to alcohol molar ratio	Microwave reaction conditions	Ester conversion (%)	Ref.
Castor	50% H_2SO_4/C	5	MeOH	1:12	MAS-1 Shanghai Sineo MW 65°C, 60 min	94	(Yuan et al., 2009)
Castor	SiO_2/50%H_2SO_4 SiO_2/50%H_2SO_4 Al_2O_3/50%KOH	10 10 10	MeOH EtOH MeOH	1:6	Domestic MW 540W 60°C, 30 min 60°C, 20 min 60°C, 5 min	95 95 95	(Perin et al., 2008)
Jatropha	KOH	1.5	MeOH	1:7.5	Start Synth-Milestone 1200W 65°C, 2 min	99	(Shakinaz et al., 2010)
Waste frying	SrO $Sr(OH)_2$	1.5	MeOH	1:4 wt	Domestic MW 900W 60°C, 40 s	99 97	(Koberg et al., 2011)

Pongamia Pinnata	NaOH KOH	0.5 1.0	MeOH	1:6	Start Synth-Milestone 1200W 60°C, 7 min	95.3 96.0	(Kumar et al., 2011)
Rapeseed	NaOH KOH	1.0	MeOH	-	Start Synth-Milestone 1200W 60°C, 5 min	91.7 90.8	(Azcan and Danisman, 2008)
Soybean	Nano CaO	3.0	MeOH	1:7	ETHOS900 Milestone 900W 65°C, 60 min	96.6	(Hsiao et al., 2011)
Soybean	Novozym 435	3.0	MeOH	1:6	MCR-3 Shanghai JieSi 800W 40°C, 12h	94.0	Yu et al., 2010)
Canola	ZnO/La$_2$O$_2$CO$_3$	1.0	MeOH	1:1 wt	Biotage MW reactor. <100°C, 5 min	>95	(Jin et al., 2011)
Camelina	BaO SrO	1.5 2.0	MeOH	1:9 1:12	Domestic MW 800W 100°C, 4 min 60°C, 4 min	95 78	(Patil et al., 2011)
Camelina	NaOH KOH BaO SrO BaCl$_2$/ AA SrCl$_2$/ AA	0.5 1.0 1.5 2.0 2.0 2.0	MeOH	1:9	Domestic MW 800W 60°C, 60 s 60°C, 60 s 60°C, 4 min 60°C, 4 min 60°C, 5 min 60°C, 5 min	95 85 95 95 27 20	(Patil et al., 2010)
Safflower	NaOH	1.0	MeOH	1:10	Start labstation-Milestone 60°C, 6 min	98.4	(Duz et al., 2011)
Soybean Rice Bran	NaOH	0.6	EtOH	1:5	ETHOS E-Milestone 73°C, 10 min	99.25 99.34	(Terigar et al., 2010)
Rapeseed	KOH	1.0	ButOH	1:4	MARS CEM Corp. 117°C, 30 min	100	(Geuens et al., 2008)
Soybean	H$_2$SO$_4$ KOH	5.0 1.0	ButOH ButOH	1:6	CEM Discover 300W CEM MARS 1600W 100°C, 15 min 120°C, 1 min	93 93	(Leadbeater et al., 2008)
Jatropha Waste frying	NaOH	1.0	MeOH	1:12	MW650 Aurora Instruments MW discovery 65°C, 7 min	89.7 88.63	(Yaakob et al., 2009)
Palm	H$_2$SO$_4$	3.0	MeOH IsoprOH IsoBuOH 2-BuOH IsopentOH	1:30	Domestic MW 1000W 60°C, 5h 75°C, 5h 105°C, 5h 90°C, 5h 115°C, 5h	49.40 62.39 67.39 62.39 75.00	(Mazo and Rios, 2010a)

Palm	NaOCH₃	0.9	MeOH IsoprOH IsoBuOH 2-BuOH IsopentOH	1:27	Domestic MW 1000W 60°C, 1h 75°C, 1h 105°C, 1h 90°C, 1h 115°C, 1h	99.9 99.87 88.39 83.19 81.63	(Mazo and Rios, 2010a)
Palm	K₂CO₃	3.0	MeOH IsoprOH IsoBuOH 2-BuOH IsopentOH	1:20	Domestic MW 1000W 60°C, 3h 75°C, 3h 105°C, 3h 90°C, 3h 115°C, 3h	8.63 49.51 67.59 52.00 54.59	(Mazo and Rios, 2010b)
Palm	KOH	1.5	EtOH	1:4	Domestic MW 800W 70°C, 5 min	97.4	(Suppalkpanya et al., 2010a)

Table 2. Microwave assisted transesterification.

2.3 Optimization production biodiesel under MW irradiation

Some examples about the obtaining of biodiesel making a response surface methodology (RSM) was used to analyze the influence of the process variables (oil to methanol ratio, catalyst concentration, and reaction time) on the fatty acid methyl ester conversion, are shown in Table 3, where is confirmed that the microwave energy has a significant effect on esterification and transesterification reactions.

Oil	Catalyst	Catalyst amount (%wt)	Alcohol	Oil to alcohol molar ratio	Microwave reaction conditions	Ester conversion (%)	Ref.
Dry algae	KOH	2.0	MeOH	1:12 (wt/vol)	Domestic MW 800W 60°C, 4min	64.18	(Patil et al., 2011)
Macauba	Novozyme 435 Lipozyme IM	2.5	EtOH	1:9	Synthos 3000-Anton Paar. 1400W. 30°C, 15min	45.2 22.9	(Nogueira et al., 2010)
Pongamia pinnata	Esterification: H₂SO₄ Transesterification: KOH	3.73 1.33	MeOH MeOH	33.83% (w/w) 33.4% (w/w)	Domestic MW 800W 60°C, 190s 60°C, 150s	87.4 89.9	(Venkatesh et al., 2011)

Table 3. Recent examples of optimization of reaction conditions a for production of biodiesel from various feedstocks using response surface methodology

3. Fatty acid alkylesters production assisted by radio frequency

Radio waves, whose wavelengths range from more than 104 m to about 0.1 m, are the result of charges accelerating through conducting wires. They are generated by such electronic devices as LC oscillators and are used in radio and television communication systems (Serway and Jewett, 2004).

Radio frequency (RF) heating is a promising dielectric heating technology which provides fast heat generation through a direct interaction between an RF electromagnetic field and the object being heated (Piyasena et al., 2003). Compared to microwave heating, a popular dielectric heating technology, RF heating systems are simpler to configure and have a higher conversion efficiency of electricity to electromagnetic power (Wang et al., 2003). Moreover, RF energy has deeper penetration into a wide array of materials than microwave energy, increasing feasibility of RF heating for industrial scale applications.

Very few publications have been obtained by this alternative heating method, which use a RF heating apparatus (SO6B; Strayfield Fastran, UK). The distance between the two electrodes was fixed at 15 cm. A 150-mL conical flask coupled with a water-cooling reflux condenser was used as a reactor. Schematic diagram and photograph are shown in Fig. 1.

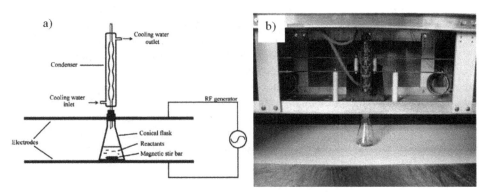

Fig. 1. a) Schematic diagram of RF heating apparatus (Lui et al., 2010) and b) Photograph of RF heating apparatus (Lui et al., 2008).

Applications to obtaining biodiese using different oils, reaction conditions and catalysts are described below:

3.1 Esterification reactions assisted by RF

Efficient biodiesel conversion from waste cooking oil with high free fatty acids (FFAs) was achieved via a two-stage procedure (an acid-catalyzed esterification followed by an alkali-catalyzed transesterification) assisted by radio frequency (RF) heating. In the first stage, with only 8-min RF heating the acid number of the waste cooking oil was reduced from 68.2 to 1.64 mg KOH/g by reacting with 3.0% H_2SO_4 (w/w, based on oil) and 0.8:1 methanol (weight ratio to waste oil). Then, in the second stage, the esterification product (primarily consisting of triglycerides and fatty acid methyl esters) reacted with 0.91% NaOH (w/w, based on triglycerides) and 14.2:1 methanol (molar ratio to triglycerides) under RF heating for 5 min, and an overall conversion rate of 98.8 ± 0.1% was achieved. Response surface methodology was employed to evaluate the effects of RF heating time, H_2SO_4 dose and methanol/oil weight ratio on the acid-catalyzed esterification. A significant positive interaction between RF heating time and H_2SO_4 concentration on the esterification was observed (Lui et al., 2010).

3.2 Transesterification reactions assisted by RF

Efficient biodiesel production from beef tallow was achieved with radio frequency (RF) heating. A conversion rate of 96.3% was obtained with a NaOH concentration of 0.6% (based on tallow), an RF heating for 5 min, and a methanol/tallow molar ratio of 9:1. Response surface methodology was employed to evaluate the influence of NaOH dose, RF heating time, and methanol/tallow ratio. The alkaline concentration showed the largest positive impact on the conversion rate. Similar fast conversion from canola oil to biodiesel was achieved in our previous work, indicating that RF heating, as an accelerating technique for biodiesel production, had a large applying area (Lui et al., 2011).

3.3 Optimization production biodiesel under RF irradiation

Fast transesterification of canola oil and methanol for biodiesel production was achieved using radio frequency (RF) heating. The conversion rate of oil to biodiesel reached 97.3% with RF heating for 3 min, a NaOH concentration (based on oil) of 1.0%, and a methanol/oil molar ratio of 9:1. A central composite design (CCD) and response surface methodology (RSM) were employed to evaluate the impact of RF heating time, NaOH concentration, and molar ratio of methanol to oil on conversion efficiency. Experimental results showed that the three factors all significantly affected the conversion rate. NaOH concentration had the largest influence, with the effect being more pronounced at lower (0.2-0.6%, based on weight of oil) concentration. No evident interaction among the three factors was observed. RF heating efficiency was primarily related to the amount of NaOH and methanol. The scale of the experiment was increased by five times (from 20 to 100 g oil per batch) without decrease of the conversion rate, indicating the scale-up potential of RF heating for biodiesel production (Lui et al., 2008).

4. Fatty acid alkylesters production assisted by ultrasound

Sound waves are the most common example of longitudinal waves. They travel through any material medium with a speed that depends on the properties of the medium. As the waves travel through air, the elements of air vibrate to produce changes in density and pressure along the direction of motion of the wave. If the source of the sound waves vibrates sinusoidally, the pressure variations are also sinusoidal (Serway and Jewett, 2004).

Sound waves are divided into three categories that cover different frequency ranges (1) Audible waves lie within the range of sensitivity of the human ear. They can be generated in a variety of ways, such as by musical instruments, human voices, or loudspeakers. (2) Infrasonic waves have frequencies below the audible range. Elephants can use infrasonic waves to communicate with each other, even when separated by many kilometers. (3) Ultrasonic waves have frequencies above the audible range. You may have used a "silent" whistle to retrieve your dog. The ultrasonic sound it emits is easily heard by dogs, although humans cannot detect it at all. Ultrasonic waves are also used in medical imaging (Serway and Jewett, 2004).

Sonochemistry is a branch of chemical research dealing with the chemical effects and applications of ultrasonic waves, that is, sound with frequencies above 20 kHz that lie beyond the upper limit of human hearing. The development of ultrasound in organic synthesis began on 1930 when Richards and Loomis, 1927, applied ultrasound (100-500 KHz) in organic synthesis for determine the effect on the solubility of gases for first time. Developments were very slow, then Luche and Damiano, 1980, reported metal activation reactions using ultrasound probes. Thereafter, reaction systems using US to speed up chemical reactions have been developed.

A low frequency ultrasonic irradiation could be useful for transesterification of triglyceride with alcohol. Ultrasonication provides the mechanical energy for mixing and the required activation energy for initiating the transesterification reaction (Singh et al., 2007). Ultrasonication increases the chemical reaction speed and yield of the transesterification of vegetable oils and animal fats into biodiesel. Ultrasonic assisted transesterification method presents advantages such as shorter reaction time and less energy consumption than the conventional mechanical stirring method, efficient molar ratio of methanol to TG, and simplicity (Ji et al., 2006; Siatis et al., 2006).

Many researchers have tried to solve the mass-transfer limitation problem in biodiesel synthesis using ultrasonic cavitation and hydrodynamic cavitation. Cavitation has been shown to efficiently speed up the transesterification reaction because it simultaneously supplies heating as well as the stirring effect as a result of jet formation on bubble collapse. Cavitation is basically the formation, growth, and implosive collapse of gas or vapour filled microbubbles and can be induced acoustically (using ultrasound) or hydrodynamically in a body of liquid. The collapse of these bubbles lead to local transient high temperatures (g 5000 K) and pressures (g 1000 atm), resulting in the generation of highly reactive species, such as $OH\bullet$, $HO_2\bullet$, and $H\bullet$ radicals in water. Cavitation effects also increase the mass and heat transfers in a medium and accelerate the reaction rates and yields (Mahamuni and Adewuyi, 2009).

Main factors that vary the yielding in the production of biodiesel using US are:

Effect of Ultrasonic Frequency on Biodiesel Yield. The frequency of the ultrasound has a significant effect on the cavitation process because it alters the critical size of the cavitation bubble, which in turn changes the intensity of the collapse of the cavitation bubbles.

Effect of Ultrasonic Power on Biodiesel Yield. It is well-known that as the ultrasonic power increases, the size of the cavitation bubbles increase leading to more intense collapse of bubble, which causes better emulsion formation of oil and methanol resulting into higher interfacial surface area for mass transfer and hence the higher biodiesel yield. The BD yield increased with increasing ultrasonic power from 150 to 450 W, but the ME content decreased at ultrasonic powers over 450 W. This is due to the decrease of the real irradiation time caused by the increase in the pulse interval required for tuning the temperature due to the extension of the irradiation power (Lee et al., 2011).

Effect of Catalyst Loading. As the amount of KOH increases, the concentration of methoxide anions, which are responsible for nucleophilic attack on the triglyceride molecules to produce biodiesel, also increase, resulting in higher biodiesel yield.

Effect of Oil/Methanol Molar Ratio. As oil and methanol are not miscible into each other, they form a heterogeneous reaction mixture and mass transfer between these two phases becomes important for the transesterification reaction. The presence of ultrasound can help increase the mass transfer between the two phases by the formation of a fine emulsion, which increases the interfacial area between the two phases. Ultrasound can also increase the mass transfer coefficient due to the presence of acoustic streaming and jet formations at the end of cavitation bubble collapse near the phase boundary between oil and methanol phases.

As shown in Fig. 2, the factors with more contribution to the production of biodiesel are ultrasonic power and catalyst loading, then oil/methanol molar ratio and finally, the frequency.

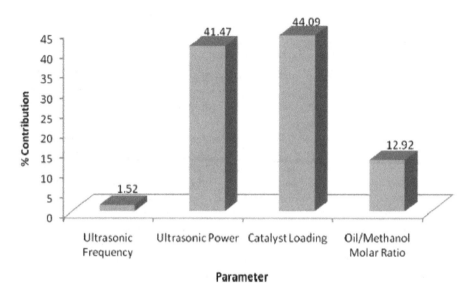

Fig. 2. Percentage contribution of the factors contributing to the production of biodiesel by ultrasound

4.1 Esterification reactions assisted by ultrasound
Not found reports of this methodology for the esterification reaction in the obtaining of biodiesel.

4.2 Transesterification reactions assisted by ultrasound
The works that use US in the transesterificatication reaction for the obtaining biodiesel use edible oils as soybean, triolein, palm, canola, fish and coconut. Also, use no edible oil as Jatropha, homogeneous basic catalysts as KOH, and heterogeneous basic catalysts as CaO, SrO, BaO, Na/SiO$_2$ and Novozym435 enzymes and lipase. Table 8 shows the work carried out for bio-diesel production from various feedstocks under different conditions using ultrasound irradiation.

The equipments used are conformed by transducer, cleaner and probe used in batch processes. In recent years, chemistry in flowing systems has become more prominent as a method of carrying out chemical transformations, ranging in scale from microchemistry up to kilogram-scale processes. Compared to classic batch ultrasound reactors, flow reactors stand out for their greater efficiency and flexibility as well as lower energy consumption. Cintas et al., 2010, developed a new ultrasonic flow reactor, a pilot system well suited for reaction scale up. This was applied to the transesterification of soybean oil with methanol for biodiesel production. This reaction is mass-transfer-limited initially because the two reactants are immiscible with each other, then because the glycerol phase separates together with most of the catalyst (Na or K methoxide). In our reactor a mixture of oil (1.6 L), methanol and sodium methoxide 30% in methanol (wt/wt ratio 80:19.5:0.5, respectively) was fully transesterified at about 45 C in 1 h (21.5 kHz, 600 W, flow rate 55 mL/min). The same result could be achieved together with a considerable reduction in energy consumption, by a two-step procedure: first a conventional heating under mechanical stirring (30 min at 45°C), followed by ultrasound irradiation at the same temperature (35 min, 600 W, flow rate 55 mL/min). Our studies confirmed that high-throughput ultrasound applications definitively require flow reactors (Cintas et al., 2010). The detailed scheme of the system is showed in Fig. 3.

Oil	Catalyst	Catalyst amount (%wt)	Alcohol	Oil to Alcohol molar ratio	Ultrasonic reaction condition	Source of ultrasound	Ester conversion (%)	Ref
Soybean	KOH	0.5	MeOH	1:6	611kHz, 139W, 26°C, 30min	Multifrequency transducer UES300C sonochemist	90	(Mahamuni and Adewuyi, 2009)
Triolein	KOH	1.0	MeOH	1:6	40kHz, 1200W, 25°C, 30min	Honda electronic cleaner	99	(Hanh et al., 2008)
Soybean	Novozym 435	6.0	MeOH	1:6	40kHz, 500W (50%), 40°C, 4h 0.5%v/v tert-amyl alcohol/oil	Ultrasonic bath KQ500DV Kunshan	96	(Yu et al., 2010)
Soybean	NaOCH3	20g (30% in MeOH)	MeOH	1.6L:80g MeOH	21.5kHz,600W, 45°C, 1h Flow 55mL/min	3 transducer (21.5kHz)	90	(Cintas et al., 2010)
Jatropha	Na/SiO₂	3.0	MeOH	1:9	24kHz, 200W, 15min	UP200S Hielscher ultrasonic Gmblt	98.53	(Kumar et al., 2010a)
Canola Soybean Corn	KOH	1.0	MeOH	1:6	450W, 55°C, 30min	Probe type VCX-600	98 97 95	(Lee et al., 2011)
Jatropha	Lipase Chromobacterium viscosum	5.0	MeOH		0.7s, 100W/m3, 30min	UP200S Hielscher ultrasonic Gmblt	84.5	(Kumar et al., 2011)
Palm	KOH	20.0	MeOH		Petroleum ether Ethyl methyl ketone 47kHz, 340W, 60°C, 2h	Water bath Bransonic cleaner	75.2 60	(Boey et al., 2011)
Palm	CaO SrO BaO	3.0	MeOH		20kHz, 200W (50%), 65°C, 60min	Transducer and probe	77.3 95.2 95.2	(Mootabadi et al., 2010)
Waste frying	KOH	First stage 0.7 Second stage 0.3	MeOH		20kHz, 25°C, 5min	Horn transducer	81 99	(Thanh et al., 2010)
Fish	NaOC₂H₅	0.8	EtOH		35kHz, 20kHz, 20°C, 30min	Bath Probe	95 95	(Armenta et al., 2007)
Coconut	KOH	0.75	EtOH		24kHz, 200W, 7min	UP200S Hielscher ultrasonic Gmblt	98	(Kumar et al., 2010b)

Table 4. Ultrasound assisted transesterification

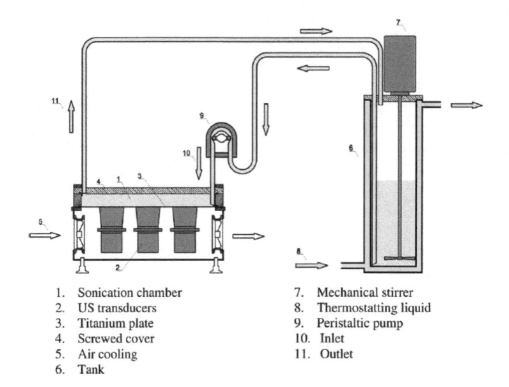

1. Sonication chamber
2. US transducers
3. Titanium plate
4. Screwed cover
5. Air cooling
6. Tank

7. Mechanical stirrer
8. Thermostatting liquid
9. Peristaltic pump
10. Inlet
11. Outlet

Fig. 3. Detailed scheme of the system for biodiesel production (Cintas et al., 2010).

4.3 Optimization production biodiesel under ultrasound

This paper utilizes the Taguchi optimization methodology (L9 orthogonal array) to optimize various parameters for the ultrasound-assisted, KOH-catalyzed transesterification of soybean oil with methanol. The statistical tool used in the Taguchi method to analyze the results is the analysis of variance (ANOVA). The optimum conditions are determined to be 581 kHz, 143 W, 0.75% (w/w) KOH loading at 1:6 oil/methanol molar ratio, resulting in more than 92.5% biodiesel yield in less than 30 min. Confirmation experiments have been performed to prove the effectiveness of the Taguchi technique after the optimum levels of process parameters are determined. (Mahamuni et al., 2010).

5. Future development

Most reports are aimed to the obtaining of biofuels of first-generation. However, these methodologies of synthesis are directed to the obtaining of fuels of second and third generation to promote sustainable chemistry and the use of renewable raw materials that not compete with foods.

The coupling of biotechnological processes with these new technologies would allow the improvement of existing processes, reducing time and increasing the production.

The development of additives that improve the properties of biodiesel would allow an improvement in cold flow properties for biodiesel from oils as palm, which can be obtained with these technologies.
Efforts must concentrate on developing these technologies at pilot and industrial levels, with continuous processes, low energy consumption, economic and insurance.

6. References

Armenta, R.E.; Vinatoru, M.; Burja, A.M.; Kralovec, J.A.; Barrow, C.J. (2007) Fish Oil Transesterification of Fish Oil to Produce Fatty Acid Ethyl Esters Using Ultrasonic Energy. J Am Oil Chem Soc, Vol.84, No.11, (November 2007), pp.1045–1052 ISSN 0003-021X

Azcan, N. & Danisman, A. Microwave assisted transesterification of rapeseed oil. Fuel, Vol.87, No. 10-11, (August 2008), pp.1781–1788. ISSN: 0016-2361

Bandow, H. (2010) A two-step continuous ultrasound assisted production of biodiesel fuel from waste cooking oils: A practical and economical approach to produce high quality biodiesel fuel. Bioresource Technology, Vol.101, No.14, (July 2010), pp.5394–5401, ISSN 0960-8524

Barnard, T.M.; Leadbeater, N.E.; Boucher, M.B.; Stencel, L.M. & Wilhite, B.A.(2007) Continuous-Flow Preparation of Biodiesel Using Microwave Heating. Energy & Fuels, Vol.21, No.3, pp.1777-1781, ISSN: 0887-0624

Boey, P.L; Ganesan, S.; Maniam, G.P. & Ali, D.M.H. (2011) Ultrasound aided in situ transesterification of crude palm oil adsorbed on spent bleaching clay. Energy Conversion and Management, Vol.52, No.5, (May 2011), pp.2081–2084, ISSN: 0196-8904

Cintas, P.; Mantegna, S.; Calcio, E. & Cravotto, G. (2010) A new pilot flow reactor for high-intensity ultrasound irradiation. Application to the synthesis of biodiesel. Ultrasonics Sonochemistry, Vol.17, No.6 (August 2010) 985–989. ISSN: 1350-4177

Dharmendra Kumar, Gajendra Kumar, Poonam, C.P. Singh. (2010) Fast, easy ethanolysis of coconut oil for biodiesel production assisted by ultrasonication. Ultrasonics Sonochemistry, Vol.17, No.3, (March 2010), pp. 555–559. ISSN: 1350-4177

Duz, M.Z.; Saydut, A. & Ozturk, G. (2011) Alkali catalyzed transesterification of safflower seed oil assisted by microwave irradiation. Fuel Processing Technology, Vol.92, No.3, (March 2011), pp.308–313, ISSN: 0378-3820

Fangrui, Ma.; & Milford, A. (1999) Biodiesel Production: a Review. Bioresource Technology, Vol.70, No.1, (October 1999), pp. 1-15, ISSN 0960-8524

Gedye, R.; Smith, F.; Westaway K.; Ali, H.; Baldisera, L.; Laberge, L. & Rousell, J. (1986) The use of microwave ovens for rapid organic synthesis. Tetrahedron Lett, Vol.27, No.3, pp.279–82, ISSN: 0040-4039

Geuens, J.; Kremsner, J.M.; Nebel, B.A.; Schober, S.; Dommisse, R.A.; Mittelbach, M.; Tavernier, S.; Kappe, O.; & Maes, B. (2008) Microwave-Assisted Catalyst-Free Transesterification of Triglycerides with 1-Butanol under Supercritical Conditions. Energy & Fuels, Vol.22, No.1, pp.643–645, ISSN: 0887-0624

Giguere, R.J.; Bray, T.L.; Duncan, S.M. & Majetich, G. (1986) Application of commercial microwave ovens to organic synthesis. Tetrahedron Lett, Vol.27, No.41, pp.4945–58, ISSN: 0040-4039

Groisman, Y. & Gedanken, A. (2008) Continuous flow, circulating microwave system and its application in nanoparticle fabrication and biodiesel synthesis. J Phys Chem C, Vol.112, No.24, pp.8802-8808, ISSN: 1932-7447.

Hanh, H.D.; The Dong, N.; Starvarache, C.; Okitsu, K.; Maeda, Y. & Nishimura, R. (2008) Methanolysis of triolein by low frequency ultrasonic irradiation. Energy Conversion and Management, Vol.49, No.2, (February 2008), pp. 276–280, ISSN: 0196-8904

Harvey, B.G & Meylemans, H.A. (2011) The role of butanol in the development of sustainable fuel technologies. J Chem Technol Biotechnol, Vol.86, No.1, (January 2011), pp.2–9, ISSN: 0268-2575

Hideki, F.; Akihiko, K. & Hideo, N. (2001) Biodiesel fuel production by transesterification of oils. J Biosci Bioeng, Vol.92, No.5, pp.405-416, ISSN: 1389-1723

Hsiao, M.C; Lin, C.C. & Chang, Y.H. Microwave irradiation-assisted transesterification of soybean oil to biodiesel catalyzed by nanopowder calcium oxide. Fuel, Vol.90, No.3, (March 2011), pp.1963–1967, ISSN: 0016-2361

Jianbing, J.; Jianli, W.; Yongchao, L.; Yunliang, Y. & Zhichao, X. (2006) Preparation of biodiesel with the help of ultrasonic and hydrodynamic cavitation. Ultrasonics, Vol.44, No.1, (December 2006), pp.e411-e414, ISSN: 0041-624X

Jin, L.; Zhang, Y.; Dombrowski, J.P.; Chen, C.; Pravatas, A.; Xu,L.; Perkins, C. & Suib, S.L. (2011) ZnO/La2O2CO3 layered composite: A new heterogeneous catalyst for the efficient ultra-fast microwave biofuel production. Applied Catalysis B: Environmental. In press. ISSN: 0926-3373

Kappe CO. (2004) Controlled microwave heating in modern organic synthesis. Angewandte Chemie International Edition, Vol.43, No.46, (November 2004), pp.6250–84, ISSN: 1521-3773

Kim, D.; Choi, J.; Kim, G.J.; Seol, S.K.; Ha, Y.C.; Vijayan, M.; Jung, S.; Kim, B.H.; Lee, D. & Park, S.S. (2011) Microwave-accelerated energy-efficient esterification of free fatty acid with a heterogeneous catalyst. Bioresource Technology, Vol.102, No.3, (February 2011), pp. 3639–3641, ISSN 0960-8524

Kim, D.; Choi, J.; Kim, G.; Seol, S.K. & Jung, S. (2011) Accelerated esterification of free fatty acid using pulsed microwaves. Bioresource Technology, In press, ISSN 0960-8524

Koberg, M.; Abu-Much, R. & Gedanken, A. (2011) Optimization of biodiesel production from soybean and wastes of cooked oil: Combining dielectric microwave irradiation and a SrO catalyst. Bioresource Technology, Vol.102, No.2, (January 2011), pp. 1073–1078, ISSN 0960-8524

Kumar, S.A.; Sandun, F. & Hernandez, R. (2007) Base-catalyzed fast transesterification of soybean oil using ultrasonication. Energy & Fuel, Vol.21, No.2, pp.1161-1164, ISSN: 0887-0624

Kumar, R.; Kumar, G.R. & Chandrashekar, N. (2011) Microwave assisted alkali-catalyzed transesterification of Pongamia pinnata seed oil for biodiesel production. Bioresource Technology, Vol.102, No.11, (June 2011), pp. 6617-6620, ISSN 0960-8524

Kumar, D.; Kumar, G.; Poonam, & Singh, P. (2010) Ultrasonic-assisted transesterification of Jatropha curcus oil using solid catalyst, Na/SiO2. Ultrasonics Sonochemistry, Vol.17, No.5, (June 2010), pp.839–844, ISSN: 1350-4177

Kumar, G.; Kumar, D.; Poonam, Johari, R. & Singh, C.P. (2011) Enzymatic transesterification of Jatropha curcas oil assisted by ultrasonication. Ultrasonics Sonochemistry, In press, ISSN: 1350-4177

Lam, M.K.; Lee, K.T. & Mohamed, A.R. (2010) Homogeneous, heterogeneous and enzymatic catalysis for transesterification of high free fatty acid oil (waste cooking oil) to biodiesel: A review. Biotechnology Advances, Vol.28, No.4 , (July-August 2010), pp. 500–518, ISSN 0734-9750

Leadbeater, N.E.; Barnard, M.; & Stencel, M.L. (2008) Batch and Continuous-Flow Preparation of Biodiesel Derived from Butanol and Facilitated by Microwave Heating. Energy & Fuels, Vol.22, No.3, pp.2005–2008, ISSN: 0887-0624

Leadbeater, N.E. & Stencel, L.M. (2006) Fast, Easy Preparation of Biodiesel Using Microwave Heating. Energy & Fuels, Vol.20, No.5, pp.2281-2283, ISSN: 0887-0624

Lee, S.B.; Lee, J.D. & Hong, K. (2011) Ultrasonic energy effect on vegetable oil based biodiesel synthetic process. Journal of Industrial and Engineering Chemistry, Vol.17, No.1, (January 2011), pp.138–143 ISSN: 1226-086X

Lee, I.; Johnson, L.A. & Hammond, E. (1995) Use of branched-chain esters to reduce the crystallization temperature of biodiesel. J Am Oil Chem Soc, Vol.72, No.10, (October 1995), pp.1155-1160 ISSN 0003-021X

Lin, L.; Cunshan, Z.; Vittayapadung, S.; Xiangqian, S. & Mingdong, D. (2011) Opportunities and challenges for biodiesel fuel. Applied Energy, Vol.88, No.6, (June 2011), pp.1020–1031. ISSN: 0306-2619

Liu, S.; Wang, Y.; McDonald, T. & Taylor, S.E. (2008) Efficient Production of Biodiesel Using Radio Frequency Heating. Energy & Fuels, Vol.22, No.3, pp.2116–2120, ISSN: 0887-0624

Liu, S.; McDonald, T. & Wang, Y. (2010) Producing biodiesel from high free fatty acids waste cooking oil assisted by radio frequency heating. Fuel, Vol.89, No.10, (October 2010), pp.2735–2740 ISSN: 0016-2361

Liu, S.; Wang, Y.; Oh, J-H. & Herring, J.L. (2011) Fast biodiesel production from beef tallow with radio frequency heating. Renewable Energy, Vol.36, No.3, (March 2011), pp.1003-1007, ISSN: 0960-1481

Loupy, A. (2002) Microwave in organic synthesis, Wiley- VCH Verlag GmbH & Co. KGaA. ISBN:3-527-30514-9, Weinheim, Germany.

Luche, J.L. and Damiano, J.C. (1980) Ultrasounds in Organic Syntheses. 1. Effect on the Formation of Lithium Organometallic Reagents. J. Am. Chem. Soc, Vol.102, No.27, pp.7926-7927, ISSN: 0002-7863

Mahamuni, N.N. & Adewuyi, Y.G. (2010) Application of Taguchi Method to Investigate the Effects of Process Parameters on the Transesterification of Soybean Oil Using High Frequency Ultrasound. Energy & Fuels, Vol.24, No.3, pp.2120–2126, ISSN: 0887-0624

Mahamuni, N.N & Adewuyi, Y.G. (2009) Optimization of the Synthesis of Biodiesel via Ultrasound-Enhanced Base-Catalyzed Transesterification of Soybean Oil Using a Multifrequency Ultrasonic Reactor. Energy & Fuels, Vol.23, No.5, pp.2757–2766, ISSN: 0887-0624

Marchetti, J.; Miguel, V. & Errazua, A. (2007) Possible methods for biodiesel production. Renewable and Sustainable Energy Reviews, Vol.11, No.6, (August 2007), pp.1300–1311. ISSN: 1364-0321

Mazo, P., Rios, L. (2010) Esterification and transesterification assisted by microwave of crude palm oil. Homogeneous catalysis. Latin American Applied Research, Vol.40, No.4, p 337-342, ISSN 0327-0793

Mazo, P., Rios, L. (2010) Esterification and transesterification assisted by microwave of crude palm oil. Heterogeneous catalysis. Latin American Applied Research, Vol.40, No.4, p 343-349, ISSN 0327-0793

Melo, C.A.; Albuquerque, C.; Carneiro, J.; Dariva, C.; Fortuny, M.; Santos, A.; Egues, S. & Ramos, A. (2010) Solid-Acid-Catalyzed Esterification of Oleic Acid Assisted by Microwave Heating. Ind. Eng. Chem. Res, Vol.49, No.23, pp.12135–12139, ISSN: 0888-5885.

Melo, C.A.; Albuquerque, C.E.R.; Fortuny, M.; Dariva, C.; Egues, S.; Santos, A.F. & Ramos, A.L.D. (2009) Use of Microwave Irradiation in the Noncatalytic Esterification of C18 Fatty Acids. Energy & Fuels, Vol.23, No.1, pp.580–585, ISSN: 0887-0624

Mootabadi, H.; Salamatinia, B.; Bhatia, S. & Abdullah, A.Z. (2010) Ultrasonic-assisted biodiesel production process from palm oil using alkaline earth metal oxides as the heterogeneous catalysts. Fuel, Vol.89, No.8, (August 2010), pp.1818–1825, ISSN: 0016-2361

Mustafa, B. (2011) Potential alternatives to edible oils for biodiesel production – A review of current work. Energy Conversion and Management, Vol.52, No.2, (February 2011), pp.1479–1492, ISSN: 0196-8904

Nogueira, B.M; Carretoni, C.; Cruza, R.; Freitas, S.; Melo, P.A.; Costa-Félix,R.; Pinto, J. & Nele, M. (2010) Microwave activation of enzymatic catalysts for biodiesel production. Journal of Molecular Catalysis B: Enzymatic, Vol.67, No.1-1, (November 2010), pp.117–121, ISSN: 1381-1177

Patil,P.; Gude, V.G.; Pinappu, S. & Deng, S. (2011) Transesterification kinetics of Camelina sativa oil on metal oxide catalysts under conventional and microwave heating conditions. Chemical Engineering Journal, Vol.168, No.3, (April 2011), pp.1296–1300. ISSN: 1385-8947

Patil, P.D.; Gude, V.G.; Camacho,L.M. & Deng, S. (2010) Microwave-Assisted Catalytic Transesterification of Camelina Sativa Oil. Energy & Fuels, Vol.24, No.2, pp. 1298–1304, ISSN: 0887-0624

Patil, P.D.; Gude, V.G.; Mannarswamy, A.; Cooke, P.; Munson-McGee, S.; Nirmalakhandan, N.; Lammers, P. & Deng, S. (2011) Optimization of microwave-assisted transesterification of dry algal biomass using response surface methodology. Bioresource Technology, Vol.102, No.2, (January 2011), pp.1399–1405, ISSN 0960-8524

Perin, G.; Alvaro, G.; Westphal, E.; Viana, L.H.; Jacob, R.G.; Lenardao, E.J. & D'Oca, M.G.M. (2008) Transesterification of castor oil assisted by microwave irradiation. Fuel, Vol.87, No.12, (September 2008), pp.2838–2841, ISSN: 0016-2361

Piyasena, P.; Dussault, C.; Koutchma, T.; Ramaswamy, H.S. & Awuah, G.B. (2003) Radio frequency heating of foods: principles, applications and related properties–a review. Crit Rev Food Sci Nutr, Vol.43, No.6, pp.587–606, ISSN: 1040-8398

Raymond A. S. & Jewett, J.W. (2004) Physics for Scientists and Engineers, Thomson Brooks/Cole, 6th edition, ISBN: 0534408427, Washigton, USA.

Richards, W.T. & Loomis, A.L. (1927) The chemical effects of high frequency sound waves. J. Am. Chem. Soc, 1927, Vol.49, No.12, pp.3086-3088. ISSN: 0002-7863

Richela, A.; Laurenta, P.; Wathelet, B.; Wathelet, J-P. & Paquot, M. (2011) Current perspectives on microwave- enhanced reactions of monosaccharides promoted by heterogeneous catalysts. Catalysis Today, In press, ISSN: 0920-5861

Shakinaz A.E.S.; Refaat, A.A. & Shakinaz T.E.S. (2010) Production of biodiesel using the microwave technique. Journal of Advanced Research, Vol.1, No.4, (October 2010), pp. 309–314, ISSN: 2090-1232.

Siatis, N.G.; Kimbaris, A.C.; Pappas, C.S.; Tarantilis, P.A. & Polissiou, M.G. (2006). Improvement of biodiesel production based on the application of ultrasound: monitoring of the procedure by FTIR spectroscopy. J Am Oil Chem Soc, Vol.83, No.1, (January 2006), pp.53-57 ISSN 0003-021X

Siddiquee, M.N. & Rohani, S. (2011) Lipid extraction and biodiesel production from municipal sewage sludges: A review. Renewable and Sustainable Energy Reviews, Vol.15, No.2, (February 2011), pp.1067–1072 ISSN: 1364-0321

Socha, A. & Sello, J.K. (2010) Efficient conversion of triacylglycerols and fatty acids to biodiesel in a microwave reactor using metal triflate catalysts. Org. Biomol. Chem, Vol.8, No.20, pp.4753–4756, ISSN. 1477-0520

Suppalakpanya, K.; Ratanawilai, S.B. & Tongurai, C. (2010) Production of ethyl ester from esterified crude palm oil by microwave with dry washing by bleaching earth. Applied Energy, Vol.87, No.7, (July 2010), pp. 2356–2359, ISSN: 0306-2619

Suppalakpanya, K.; Ratanawilai, S.B. & Tongurai, C. Production of ethyl ester from crude palm oil by two-step reaction with a microwave system. Fuel, Vol.89, No.8, (August 2010), pp. 2140–2144, ISSN: 0016-2361

Terigar, B.G.; Balasubramanian, S.; Lima, M. & Boldor, D. (2010) Transesterification of Soybean and Rice Bran Oil with Ethanol in a Continuous-Flow Microwave-Assisted System: Yields, Quality, and Reaction Kinetics. Energy & Fuels, Vol.24, No.12, pp.6609–6615, ISSN: 0887-0624

Thostenson, E.T. & Chou, T.W. (1999) Microwave processing: fundamentals and applications. Composites Part A: Applied Science and Manufacturing, Vol.30, No.9, (September 1999), pp.1055–1071

Tierney, J.P. & Lidstrom, P. (2005) Microwave assisted organic synthesis, Blackwell Publishing Ltd, ISBN 0-8493-2371-1, Oxford, UK.

Venkatesh, H.; Regupathi, I. & Saidutta, M.B. Optimization of two step karanja biodiesel synthesis under microwave irradiation. Fuel Processing Technology, Vol.92, No.1, (January 2011), pp.100–105, ISSN: 0378-3820

Vicente, G.; Martínez, M. & Aracil, J. (2004) Integrated biodiesel production: a comparison of different homogeneous catalysts systems. Bioresource Technology, Vol.92, No.3, (May 2004), pp. 297-305, ISSN 0960-8524

Vyas, A.P.; Verma, J.L.; Subrahmanyam, N. A review on FAME production processes. Fuel, Vol.89, No.1, (January 2010), pp.1–9, ISSN: 0016-2361

Wang Y, Wig TD, Tang J, Hallberg LM. (2003) Dielectric properties of foods relevant to RF and microwave pasteurization and sterilization. J Food Eng, Vol.57, No.3, (May 2003), pp.257–68, ISSN: 0260-8774

Yaakob, Z.; Ong, B.H.; Kumar, M.N.S. & Kamarudin, S. K. (2009) Microwave-assisted transesterification of jatropha and waste frying palm oil. International Journal of Sustainable Energy, Vol.28, No.4, pp.195-201, ISSN: 1478-646X

Yuan, H.; Yang, B.L. & Zhu, G.L. (2009) Synthesis of Biodiesel Using Microwave Absorption
        Catalysts. Energy & Fuels, Vol.23, No.1, pp.548–552, ISSN: 0887-0624

Yu, D.; Tian, L.; Wu, H.; Wang, S.; Wang, Y.; Ma, D. & Fang, X. Ultrasonic irradiation with
        vibration for biodiesel production from soybean oil by Novozym 435. Process
        Biochemistry, Vol.45, No.4, (April 2010), pp.519–525. ISSN: 1359-5113

Yu, D.; Tian, L.; Ma, D.; Wu, H.; Wang, Z.; Wang, L. & Fang, X. (2010) Microwave-assisted
        fatty acid methyl ester production from soybean oil by Novozym 435. Green Chem,
        Vol.12, No.5, pp.844–850. ISSN: 1463-9262

Yusuf, N.N.A.N.; Kamarudin, S.K. & Yaakub, Z. (2011) Overview on the current trends in
        biodiesel production. Energy Conversion and Management, Vol.52, No7, (July
        2011), pp.2741–2751, ISSN: 0196-8904

3

Production of Biodiesel via In-Situ Supercritical Methanol Transesterification

Asnida Yanti Ani, Mohd Azlan Mohd Ishak and Khudzir Ismail
Fossil & Biomass Energy Research Group, Fuel Combustion Research Laboratory,
Universiti Teknologi MARA, Perlis,
Malaysia

1. Introduction

Most energy that the world is using is derived from unrenewable fossil fuel that has a great impact on environments (Warabi et al., 2004). The demand of fossil fuels is increasing very rapidly and it is estimated that the remaining world reserves will be exhausted by the year 2020, with the current rate of consumption. There is an urgent need to seek for an alternative fuels to substitute the diesel due to gradual depletion of world crude oil reserves. Research is, therefore oriented for alternative energy. Biomass is one of its candidates, because biomass energy has some advantageous in reproduction, cyclic and carbon neutral properties (Warabi et al., 2004). Biodiesel fuel is one example of biomass energy, and it is generally made of methyl esters of fatty acids produced by the transesterification reaction of triglycerides with methanol with the help of a catalyst (Clark et al., 1984). Alcoholysis of vegetable oils produces fatty acids alkyl esters that are excellent substitutes for conventional fossil diesel fuels (Selmi and Thomas, 1998; De et al., 1999). The viscosity of alkyl esters is nearly twice that of diesel fuel instead of 10–20 times as in the case of neat vegetable oil (Rathore and Madras, 2007).

The use of such edible oil to produce biodiesel is not feasible in view of big gap in the demand and supply of such oils in the country for dietary consumption. Increased pressure to augment the production of edible oils has also put limitations on the use of these oils for production of biodiesel (Sinha et al., 2008). Therefore, biodiesel is actually competing limited land availability with the food industry for the same oil crop. Thus, instead of arable land being utilized to grow food, it is now being used to grow fuel. This will then raise the price of edible oil making the biodiesel produced economically unfeasible as compared to petroleum-derived diesel. In order to overcome this issue, many researchers have begun searching for cheaper and non-edible oils to be used as alternative feedstock for biodiesel production (Kansedo et al., 2009). Few sources have been identified such as waste cooking oil (Wang et al., 2006; Chen et al., 2009) and oils from non-edible oil-producing plants such as *Jatropha curcas* (Heller, 1996; Herrera et al., 2006; Tiwari et al., 2007; Berchmans and Hirata, 2008; Chew, 2009), *Pongamia pinnata* (Meher et al., 2006; Naik et al., 2008; Pradhan et al., 2008), *Calophyllum inophyllum* (Sahoo et al., 2007), cottonseed (Demirbas, 2008; Qian et al., 2008; Rashid et al., 2009), rubber seeds (Ikwuagwu et al., 2000; Ramadhas et al., 2005) and tobacco seeds (Usta, 2005; Veljkovic et al., 2006). Obviously, developing nations have to focus their attention on oils of non-edible nature, which are cheaper (Sinha et al., 2008). In Malaysia, *Jatropha curcas* L. (JCL), could be utilized as a source for production of oil and can be grown in large scale on non-cropped marginal lands and waste lands.

JCL oil is obtained only after going through the following steps: collection of fruit from the trees, separation of seed from the hull, seed drying (Chew, 2009), oil pressing and filtration. Pressing oil from the kernel yields kernel cake (40-50%) and crude oil (50-60%). At present, in the majority of cases oil is generally pressed directly from the seed without separating the kernel and shell. This method produces seed cake (70-75%) and crude oil (25-30%) (Chew, 2009). Much of the un-extractable oil still remains in the seed cake; hence better ways of extracting the oils are needed. Among the extraction techniques reported in the literature include the use of Soxhlet extraction method (Castro and Ayuso, 1998; Ayuso and Castro, 1999; Szentmihalyi et al., 2002; Darcia and Castro, 2004), aqueous enzymatic oil extraction (Rosenthal et al., 1996; Sharma and Gupta, 2006; Jiang et al., 2010) and enzyme assisted three phase partitioning (Shah et al., 2004; Gaur et al., 2007). Some of these extraction methods, however, required a longer extraction time (Chew, 2009). Nowadays, many researchers (Papamichail et al., 2000; King et al., 2001; Cao and Ito 2003; Machmudah et al., 2008) turns to supercritical extraction techniques which is relatively rapid because of the low viscosities and high diffusivities associated with supercritical fluids.

Transesterification is the general term used to describe the important class of organic reactions where an ester is transformed into another ester through interchange of the alkoxy moiety. Several aspects, including the type of catalyst (alkaline, acid or enzyme), alcohol/vegetable oil molar ratio, temperature, purity of the reactants (mainly water content) and free fatty acid content have an influence on the course of the transesterification. In the conventional transesterification of fats and vegetable oils for biodiesel production, free fatty acid and water always produce negative effects, since the presence of free fatty acids and water causes soap formation, consumes catalyst and reduces catalyst effectiveness, all of which result in a low conversion (Demirbas, 2007). In addition to that, more catalyst is required to neutralize free fatty acids of oil with higher free fatty acids content (Kusdiana and Saka, 2004). Thus, the catalytic processes have a high production cost and are energy intensive. One primary problem is due to the vigorous stirring required for the mixing of the two-phase mixture of oil and alcohol. Another problem is the separation of catalyst after the reaction (Madras et al., 2004). Therefore, non-catalytic transesterification has been investigated.

Supercritical fluid extraction using polar solvent such as methanol as an extraction solvent is highly potential extraction technique to be used whereby high yield of oil can be achieved within a shorter time (Hawash et al., 2009). Further, at supercritical state, the solvent solubility increased dramatically, and the extracted oil is relatively low in impurities (Tan et al., 2009). However, there is no details on the maximum crude biodiesel yield can be obtained related to the in-situ supercritical methanol transesterification direct from the seeds.

In situ transesterification differs from the conventional reaction in the sense that the oil-bearing material contacts acidified alcohol directly instead of reacting with purified oil and alcohol. That is, extraction and transesterification of the seed powder proceed within the same process, with alcohol acts as an extracting solvent as well as esterification reagent (Fukuda et. al., 2001). In situ transesterification (Harrington and Evans, 1985; Marinkovic and Tomasevic, 1998; Kildiran et al., 1996; Hass et al., 2004), a biodiesel production method that utilizes the original agricultural products instead of purified oil as the source of triglycerides for direct transesterification, eliminates the costly hexane extraction process and works with virtually any lipid-bearing material. It could reduce the long production system associated with pre-extracted oil and maximize alkyl ester yield. The use of reagents and solvents is reduced, and the concern about waste disposal is avoided. This process reduces the cost of final product as this process has less number of unit operations. It is the best non-renewable source of energy with good environmental impact and easy recovery.

Thus, this study contributes in terms of design, development and improvement of the in-situ supercritical methanol transesterification of biodiesel production via high-pressure high-temperature batch-wise reactor system. In this study, biodiesel is generated directly from JCL seeds using methanol at different solvent critical states.

2. Materials and methods

2.1 Sample preparation
The *Jatropha curcas* L. (JCL) fruits were obtained with cooperation from the Plantation Unit of Universiti Teknologi MARA Perlis, Malaysia. JCL fruits were cleaned and de-hulled to separate the hull from the seeds. The seeds were then dried in an oven at 105 °C for 35 min (Akbar et al., 2009). The JCL seeds were ground using grinder and sieved through progressively finer screen to obtain particle sizes (d_p) of < 1 mm (Augustus et al., 2002). Sieving was accomplished by shaking the JCL powder in a Endecotts Shaker Model EFL2 for about 30 min and finally stored in a tightly-capped plastic container. The seeds need to be dried and ground in order to remove surface moisture content to obtain constant weight and weaken or rupture the cell walls to release oil for extraction, respectively (Akpan, 2006).

2.2 In-situ supercritical methanol transesterification
A batch type reactor at supercritical methanol was used for in-situ supercritical methanol transesterification of biodiesel from JCL seeds. The in-situ transesterification was carried out at temperatures and pressures ranging from 180 – 300 °C and 6 – 18 MPa, respectively. After a leak-check test, the reactor was pressurized with nitrogen to the desired pressure and heated to reaction temperature at a rate of 5 °C/min. After reaching desired temperature, the reaction was held for periods of 5 – 35 min. A JCL seeds-to-methanol ratio (1:15, 1:20, 1:30, 1:40 and 1:45 w/v) was also investigated. After each reaction, the vessel was removed from the heater and placed into a cold water bath to quench the reaction and depressurized to ambient pressure. The extracted product was discharged from the reactor and was vacuum-filtered on a Buchner funnel and the filter cake was washed with methanol. The extracted products from the in-situ transesterification were allowed to settle and separated into two phases in 500 ml separating funnel. It took about 30 min to separate into two phases, i.e., the top phase consists of the biodiesel (fatty acid methyl ester) and the lower phase consists of the glycerol and other minor components. The schematic diagram of the experimental apparatus of the batch-wise extraction system is shown in Fig. 1.

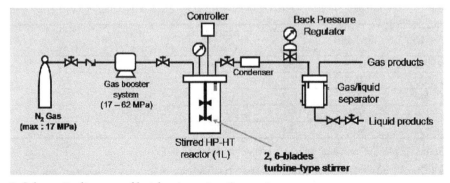

Fig. 1. Schematic diagram of batch-wise extraction system.

2.3 FAMEs analysis

The FAMEs analysis was quantified by Agilent Technologies 6890N with HP-5 5% Phenyl Methyl Siloxane capillary column (30 m by 320 μm by 0.25 mm) and a flame ionization detector. Methyl heptadecanoate (10.0 mg; internal standard) was dissolved in 1 ml hexane to prepare the standard solution. Approximately 100 mg crude methyl ester was dissolved in 1 ml standard solution for GC analysis (Hong, 2009). Approximately 1 μl sample was injected into the GC at an oven temperature of 210 °C with Helium as the carrier gas. The GC oven was programmed at 210 °C, isothermally for 15 min. the FAMEs content was calculated by use of the Equation 1:

$$C = \frac{\sum A - A_{IS}}{A_{IS}} \times \frac{C_{IS} \times V_{IS}}{m} \times 100\% \tag{1}$$

Where:

$\sum A$ = total peak area of methyl ester

A_{IS} = peak area of internal standard (methyl heptadecanoate)

C_{IS} = concentration of the internal standard solution, in mg/ml

V_{IS} = volume of the internal standard solution used, ml

m = mass of the sample, in mg

2.4 Biodiesel properties

The biodiesel was characterized by its density, viscosity, high heating value, cloud and pour points and flash points according to ASTM standards.

3. Results and discussion

3.1 Effect of temperature

The effect of temperature on percent of FAMEs yields from JCL seeds were investigated. The parameters were fixed at 12 MPa of pressure, 1:40 (w/v) of seeds-to-methanol ratio, 30 min of reaction time and at varying temperatures of 180, 200, 240, 280 and 300 °C. The results of in-situ supercritical methanol on percent of FAMEs yields from JCL seeds at various temperatures are shown in Table 1. For simplification, the data are also plotted in Fig. 2.

Temperature (°C)	Yields (%)					
	FAMEs	Methyl Palmitate	Methyl Oleate	Methyl Linoleate	Methyl Stearate	Others
180	63.9	10.3	27.9	22.1	3.6	36.1
200	76.0	13.4	34.6	23.2	4.8	24.0
240	90.3	16.2	36.4	31.1	6.6	9.7
280	97.9	18.1	39.5	33.2	7.1	2.1
300	90.9	16.3	36.6	31.3	6.7	9.1

[1]conditions: 12 MPa, 30 min and 1:40 (w/v) seeds-to-methanol ratio.

Table 1. In-situ supercritical methanol transesterification[1] results from JCL seeds at various temperatures on percent of FAMEs yield and its contents.

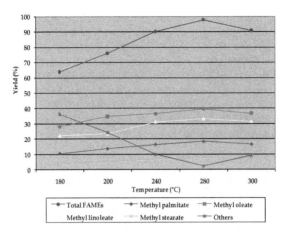

Fig. 2. In-situ supercritical methanol transesterification results from JCL seeds at various temperatures on percent of FAMEs yield and its contents.

From Table 1, the results indicate that the percent of FAMEs yields obtained at temperatures of 180 to 300 °C were in the range of 63.9 – 97.9%. The saturated FAMEs content of the seed samples are low, which is between 10.3 – 18.1% for methyl palmitate and 3.6 – 7.1% for methyl stearate. Meanwhile, the content of unsaturated FAMEs, methyl oleate and methyl linoleate are considerably higher at 27.9 – 39.5% and 22.1 – 33.2%, respectively. It should be noted that the critical temperature of methanol is at 240 °C and therefore, the conditions at 180 – 200 °C, 240 and >240 – 300 °C represent subcritical, supercritical and postcritical states of the medium, respectively.

At 180 °C, which is the lowest temperature of investigation, low yields of FAMEs (63.9%) were obtained. This observation might be due to the subcritical state of methanol or the instability of the supercritical state of methanol. It was observed that by increasing the reaction temperature to supercritical conditions had a favorable influence on the yield of ester conversion (Demirbas, 2008). Similar results have been reported by Cao et al., (2005), Madras et al., (2004) and Bunyakiat et al., (2006) on soybean oil, sunflower oil and coconut oil, respectively.

Apparently, by increasing the reaction temperature from 200 to 280 °C, the conversion increases significantly with FAMEs yields increased from 76.0 – 97.9%. The higher conversions observed in the supercritical state can be attributed to the formation of a single phase between alcohol and oil (Madras et al., 2004). Under supercritical conditions, the solubility parameter of alcohol reduces and was close to the solubility parameter of oil (Han et al., 2005). According to Petchmala et al., (2008), the increase in temperature causes the polarity of methanol to decrease, as a result of the breakdown of the hydrogen bonding of methanol, leading to an increased in the solubility of fatty acids in methanol. The complete solubility occurs as the temperature approaches the mixture critical temperature, at which point the reaction mixture became homogeneous and reaction took place rapidly. In addition, higher temperature contributed to the decomposition of cell walls, and as a result crude biodiesel yield was increased (Machmudah et al., 2007).

At 300 °C, the percent of FAMEs (90.9%) yields were slightly decreased. This observation was due to the decomposition of polyunsaturated methyl esters and unreacted triglycerides in postcritical methanol at severe high temperature (Tan et al., 2009). This finding was further supported by Xin et al., (2008) who suggested that the favorable reaction temperature adopted in supercritical methanol method should be lower than 300 °C. Reaction temperature at above 380 °C is insuitable for transesterification reaction because the oil and methyl esters tend to decompose at the highest rate. Furthermore, Kusdiana and Saka's (2001) pointed out that saturated and unsaturated FAMEs undergo side reactions such as thermal decomposition and dehydrogenation reactions at temperature >400 °C and >350 °C, respectively. In these experiments, the temperature used was lower than that of Kusdiana and Saka's work and the side reactions did not occur since the temperature was below 300 °C. Furthermore, at 300 °C, a strong burning smell of the extract was detected. Hence, at this point, there is no reason to further increase the extraction temperature beyond 280 °C.

3.2 Effects of pressure

The results of in-situ supercritical methanol transesterification on percent of FAMEs yields from JCL seeds at various pressures are shown in Table 2. For simplification, the data are also being plotted and is shown in Fig. 3. The temperature was fixed at 280 °C based on the maximized yield conditions from the previous experiment.

Pressure (MPa)	Yields (%)					
	FAMEs	Methyl Palmitate	Methyl Oleate	Methyl Linoleate	Methyl Stearate	Others
6	80.6	13.1	41.0	20.4	6.1	19.4
8	95.6	15.7	47.1	26.3	6.5	4.4
12	97.9	18.1	39.5	33.2	7.1	2.1
16	93.5	16.0	38.4	32.1	7.0	6.5
18	92.5	16.0	38.6	31.1	6.8	7.5

[1]conditions: 280 °C, 30 min and 1:40 (w/v) seeds-to-methanol ratio.

Table 2. In-situ supercritical methanol transesterification[1] results from JCL seeds at various pressures on FAMEs yield and its contents.

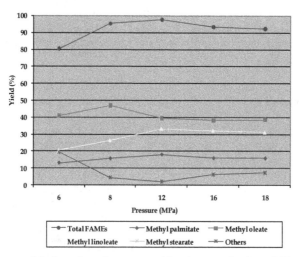

Fig. 3. In-situ supercritical methanol transesterification results from JCL seeds at various pressures on FAMEs yield and its contents.

From Table 2, the results indicated that the percent of FAMEs yields obtained at temperatures of 280 °C and pressures of 6-18 MPa was in the range of 80.6 – 97.9% with maximum yields at 12 MPa. The saturated FAMEs content of the seed samples are low, which is between 13.1 – 18.1% for methyl palmitate and 6.1 – 7.1% for methyl stearate. Meanwhile, the content unsaturated FAMEs, methyl oleate and methyl linoleate are considerably higher at 38.6 – 47.1% and 20.4 – 33.2%, respectively.

At the lowest pressure of 6 MPa, FAMEs yields are only 80.6%, but increases to 97.9% when the pressure are increased to 12 MPa. The high FAMEs yields achieved at 12 MPa, which is slightly above the critical pressure of methanol (8.09 MPa), might be due to the increase in solvent power of methanol with increasing pressure.

Further, increasing the pressure to 18 MPa, the FAMEs yield decreases slightly to 92.5%. After the pressure increased to a specific level, the increase of pressure does not cause an obvious improvement in the FAME yield (He et al., 2007). This phenomenon might be due to the maximum solubility and/or hydrogen donor ability of the solvent that has been achieved regardless of high pressure employed.

As the pressure of the system increased, the solubility parameter of the methanol decreased and is close to the solubility parameter of the oil, thus forming a single phase between the alcohol and the oil. Based on these results, it can be seen that the fact that both temperature and pressure play an important role that contributes to high extraction yield, with the later being more prominent. Based on these results, it can be seen that the fact that both temperature and pressure play an important role that contributes to high yield, with the later being more prominent.

3.3 Effects of reaction time

Table 3 and Fig. 4 shows the effect of reaction time on percent of FAMEs yields from JCL seeds using in-situ supercritical methanol transesterification. The reaction conditions were fixed based on maximum yields at optimized conditions discussed previously, i.e. 280 °C of temperature and 12.7 MPa of pressure.

Reaction time (min)	Yields (%)					
	FAMEs	Methyl Palmitate	Methyl Oleate	Methyl Linoleate	Methyl Stearate	Others
5	88.4	15.0	38.5	28.6	6.3	11.6
10	94.2	16.6	40.5	30.6	6.5	5.2
20	96.0	17.2	39.6	32.0	7.2	4.0
30	97.9	18.1	39.5	33.2	7.1	2.1
35	93.1	16.6	38.8	30.8	6.9	6.9

[1]conditions: 280 °C, 12.7 MPa and 1:40 (w/v) seeds-to-methanol ratio.

Table 3. In-situ supercritical methanol transesterification[1] results of JCL seeds at various reaction times on percent of FAMEs yield and its contents.

From Table 3 and Fig. 4, the results indicated that the percent of FAMEs yields obtained at temperatures of 280 °C, pressures of 12.7 MPa, seeds-to-methanol ratio of 1:40 (w/v) and reaction time of 5 – 35 min was in the range of 88.4 – 97.9% with maximum yields at 30 min.

The saturated FAMEs content of the seed samples are low, which is between 15.0 – 18.1% for methyl palmitate and 6.3 – 7.2% for methyl stearate. Meanwhile, the content unsaturated FAMEs, methyl oleate and methyl linoleate are considerably higher at 38.5 – 39.5% and 28.6 – 33.2%, respectively.

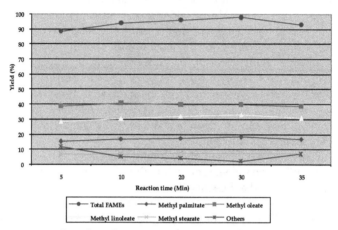

Fig. 4. In-situ supercritical methanol transesterification results from JCL seeds at various reaction times on percent of FAMEs yield and its contents.

From the results, it can be seen that the percent of FAMEs yields were only 88.4% at 5 min of reaction time. According to Saka and Kusdiana (2001), in the common method, the reaction is initially slow because of the two-phase nature of the methanol/oil system, and slows even further because of the polarity problem even with the help of an acid or an alkali catalyst. However, as described in this work, supercritical method can readily solve these problems because of the supercritical temperature and pressure employed. It can be seen that the conversion was increased in the reaction time ranges between 5 and 30 min with the percent of FAMEs yields showed a slight increase in the range of 88.4 – 97.9%.

Further, the results indicated that an extension of the reaction time from 30 to 35 min had leads to a reduction in the FAMEs yield (93.1%). This is because longer reaction time enhanced the hydrolysis of esters (reverse reaction of transesterification), resulted in loss of esters as well as causing more fatty acids to form soap (Eevera et al., 2009). Hence, for this process, there is no reason to prolong the reaction time beyond 30 min. Thus, the reaction time of 30 min can be considered as the economic reaction time by considering the percent of crude biodiesel and FAMEs yields being achieved.

3.4 Effects of seeds-to-methanol ratio

Table 4 and Fig. 5 shows the effect of seeds-to-methanol ratio on percent of FAMEs yields from JCL seeds using in-situ supercritical methanol transesterification. The reaction conditions were fixed based on maximized yields at optimized conditions discussed previously, i.e. 280 °C of temperature and 12.7 MPa and 30 min of reaction time with varying seeds-to-methanol ratio of 1:20, 1:30 and 1:40 (w/v).

Seed-to-methanol ratio (w/v)	Yields (%)					
	FAMEs	Methyl Palmitate	Methyl Oleate	Methyl Linoleate	Methyl Stearate	Others
1:15	89.0	15.2	37.4	29.8	6.6	11.0
1:20	94.4	16.9	38.8	31.6	7.1	5.6
1:30	95.9	17.5	38.6	32.3	7.5	4.1
1:40	97.9	18.1	39.5	33.2	7.1	2.1
1:45	97.0	17.4	40.1	32.3	7.2	3.0

[1] conditions: 280 °C, 12.7 MPa, 30 min reaction time.

Table 4. In-situ supercritical methanol transesterification[1] results of JCL seeds at various seed-to-methanol ratios on percent of FAMEs yield and its contents.

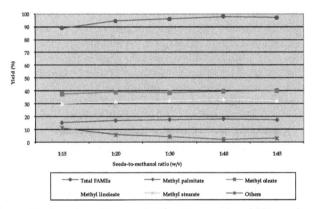

Fig. 5. In-situ supercritical methanol transesterification results of JCL seeds at various seeds-to-methanol ratios on percent of FAMEs yield and FAMEs contents.

From Table 4 and Fig. 5, the results indicated that the percent of FAMEs yields obtained at temperatures of 280 °C, pressures of 12.7 MPa, reaction time of 30 min and at various seeds-to-methanol ratio (1:15 – 1:45 w/v) was in the range of 89.0 – 97.9%, with maximum yields at 1:40 (w/v). The saturated FAMEs content of the seed samples are low, which is between 15.2 – 18.1% for methyl palmitate and 6.6 – 7.5% for methyl stearate. Meanwhile, the content unsaturated FAMEs, methyl oleate and methyl linoleate are considerably higher at 37.4 – 39.5% and 29.8 – 33.2%, respectively.

Obviously, at the lowest seeds-to-methanol ratio of 1:15 (w/v), the percent of FAMEs yields was relatively low (89.0%) and increased with increasing seeds-to-methanol ratio. When the methanol content in the supercritical fluids increased, the percent conversion of methyl ester also increased. The higher methanol content is favorable not only because more molecules of methanol surround the oil molecules but also because it contributes to the lower critical temperature of the mixture. Maximum percent of crude biodiesel and FAMEs yields were obtained at a 1:40 (w/v) of seeds-to-methanol ratio. This is a significant difference from conventional catalytic reaction for which at least 1 h of reaction time is needed to attain the same yield. In this reaction, an excess of methanol was used in order to shift the equilibrium in the direction of the products (Demirbas, 2007). Kusdiana and Saka (2001) have suggested that higher molar ratios of methanol to oil also result in a more efficient transesterification reaction. The results obtained shows good agreement with previous work, where maximum conversion was obtained for rapeseed oil (Saka and Kusdiana, 2001) at molar ratio of 42:1, for various vegetable oils (Demirbas, 2002; Diasakou et al., 1998; Ma, 1998) and linseed oil (Varma and Madras, 2007) at molar ratio of 41:1 and 40:1, respectively. According to Bunyakiat et al., (2006), when the methanol content in the supercritical fluids increased, the percent of methyl esters conversion also increased.

The higher methanol content is favorable not only because more molecules of methanol surround the oil molecules but also because it contributes to the lower critical temperature of the mixture. It can be seen that an increment in seed-to-methanol ratio can enhance biodiesel yield due to higher contact area between methanol and triglycerides. However, when the ratio is beyond 40, the yield of biodiesel begins to decrease substantially. This might be due to the restriction of the reaction equilibrium and difficulties in separating excessive methanol from methyl esters and glycerol, which subsequently lowered the yield of biodiesel (Tan et al., 2009).

Moreover, it was observed that for high seeds-to-methanol ratio added the set up required longer time for the subsequent separation stage since separation of the FAMEs layer from the organic layer becomes more difficult with the addition of a large amount of methanol. This is due to the fact that methanol, with one polar hydroxyl group, can work as an emulsifier that enhances emulsion. Operating beyond the optimal value, the ester yield would not be increased but will result in additional cost for methanol recovery (Eevera et al., 2009). Therefore, increasing the seeds-to-methanol ratio is another important parameter affecting the FAMEs yield. This report is in line with the results of many investigations based on neat vegetable oils (Freedman et al., 1984; Zhang et al., 2003; Leung et al. 2006, Eevera et al., 2009).

3.5 Biodiesel characterization

The biodiesel obtained through the one-step supercritical methanol extraction and transesterification in-situ process in this experiment was dark yellow in color. Compositions of samples were analyzed by GC. Figure 6 shows the total ion current chromatogram of the biodiesel. Furthermore, Table 5 shows the names, structure and compositions of *Jatropha curcas* L. FAMEs.

Fig. 6 depicts the gas chromatographic evaluation of the FAMEs produced over the course of reaction. The methyl esters analyzed by GC appear in the retention time of less than 15 min in the chromatograms. The weight percentages were similar for all of the variables condition; temperature, pressure, reaction time and seeds-to-methanol ratio of in-situ transesterification, as suggested by Carrapiso et al., (2000) that transesterification was random. The average saturated FAMEs content of the seed samples are low: 18.1% for methyl palmitate (C17:0) and 7.1% for methyl stearate (C19:0). The average content of the unsaturated FAMEs, methyl oleate (C19:1) and methyl linoleate (C19:2) are considerably higher at 39.5 and 33.2%, respectively which are comparable to the fatty acid composition in crude JCL oil feedstock. Depending on the origin, either oleic or linoleic acid content is higher. In this case, the seed oil belongs to the oleic or linoleic acid group, to which similar to the majority of vegetable oils (Carrapiso et al., 2000).

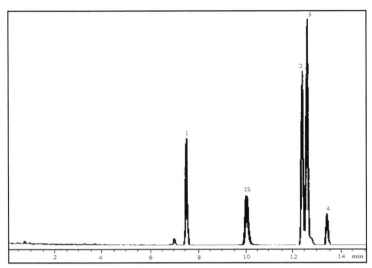

IS: Internal standard (Methyl heptadecanoate)

Fig. 6. Total ion current chromatogram of the biodiesel.

Peak No.	Name	Wt%
1	Methyl Palmitate	18.46
2	Methyl Oleate	40.41
3	Methyl Linoleate	33.91
4	Methyl stearate	7.22

Table 5. Names, structure and compositions of *Jatropha curcas* L. FAMEs.

3.5.1 Biodiesel characterization

Vegetable oil methyl esters, commonly referred to as "biodiesel" are prominent candidates as alternative Diesel fuels. Biodiesel is technically competitive with or offers technical advantages compared to conventional petroleum Diesel fuel. The vegetable oils, as alternative engine fuels, are all extremely viscous with viscosities ranging from 10 to 20 times greater than that of petroleum Diesel fuel (Demirbas, 2003). The purpose of the transesterification process is to lower the viscosity of the oil. In this study, in-situ supercritical methanol transesterification for production of biodiesel from Jatropha curcas L. (JCL) seeds was generate via 1000 ml high-temperature high-pressure batch-wise reactor system in an absence of catalyst. The reaction conditions were conducted at 280 °C of temperature, 12.7 MPa of pressure, 30 min of reaction time and 1:40 of seeds-to-methanol ratio at 450 rpm of stirring rate. Samples of the biodiesel obtained from the in-situ experiment were determined using reference methods published by American Society for Testing and Materials (ASTM) D6751. In order to ensure that it can be used in diesel engine without any modification, the properties of biodiesel produced from this in-situ transesterification reaction was comparable with fuel properties of No. 2 Diesel. Fuel properties of No. 2 Diesel, JCL biodiesel and ASTM D6751 derived biodiesel standards is shown in Table 6 for comparison.

Properties	No.2 Diesel[a]	JCL biodiesel[b]	ASTM D6751[a]	JCL Biodiesel (This study)
Specific gravity	0.85	0.86 to 0.87	0.87 to 0.90	0.87
Kinematic viscosity @ 40 °C (cSt)	1.9 to 4.1	4.23 to 5.65	1.9 to 6	5.27
Cloud point (°C)	-19 to -8	8 to 10.2	Report	-2.06
Pour point (°C)	-34 to -10	4.2 to 6	-15 to 10	0
Flash point (°C)	51 to 85	130 to 192	130 min	100
Calorific value (MJ/kg)	45.0 to 45.3	38.5-42.7	41.0	39.3

[a]Demirbas, (2008); Encinar, (2005); Vyas, (2009)
[b]Ghadge and Rehman, (2005); Vyas, (2009); Sahoo and Das, (2009)

Table 6. Fuel properties of No. 2 Diesel and JCL biodiesel.

The properties of biodiesel produced from this in-situ supercritical methanol transesterification were comparable with fuel properties of commercial No. 2 Diesel. It was found that specific gravity of JCL biodiesel was 0.87 g/cm^3 and it falls between the ASTM D6751 ranges. Fuel injection equipment operates on a volume metering system, hence a higher density for biodiesel results in the delivery of a slightly greater mass of fuel (Demirbas, 2005). The kinematic viscosity was 5.27 cSt. Among the general parameters for biodiesel the viscosity of FAMEs can go very high levels and hence it is important to control it within an acceptable level to avoid negative impacts on fuel injector's system performance (Murugesan et al., 2009). The flash point was determined to be at 100 °C. Since biodiesel has a higher flash point than diesel, it is a safer fuel than diesel. Addition of a small quantity of biodiesel with diesel increases the flash point of diesel which can result in improved fire safety for transport purpose (Lu et al., 2009) and it is safer to store biodiesel-diesel blends in comparison to diesel alone (Sahoo et al., 2009). Meanwhile, the pour point was measured to be 0 °C which was slightly higher than that of No. 2 Diesel fuel. This might be due to the presence of wax, which begins to crystallize with the decrease in temperature. This finding was agreed with Vyas et al., (2009) and Raheman and Ghadge, (2007). The problems of higher pour point of JCL biodiesel could be overcome by blending with diesel. The cloud point was reported to be -2.06 °C. The cloud point depends upon the feedstock used and must be taken into consideration if the fuel is to be used in cold environments (Fernando et al., 2007). The calorific value of JCL biodiesel was 39.3 MJ/kg, which was almost 88% of the calorific value of diesel (44.8 MJ/kg). The lower calorific value of JCL is because of the presence of oxygen in the molecular structure, which is confirmed by elemental analysis also. Furthermore, the presence of oxygen in the biodiesel helps for complete combustion of fuel in the engine. These findings were also agreed by Sinha et al., (2008). Therefore, they could be excellent substitutes and blends of No. 2 diesel fuel.

4. Conclusions

Based on the findings, it can be concluded that temperature is an important property in this in-situ process. As the temperature increased, the crude biodiesel and FAMEs yields also increased. The crude biodiesel and FAMEs of the yields reached a maximum (59.9 and 97.9, respectively) at 280 °C and then decreased with increasing temperature. The loss was caused by thermal decomposition, dehydrogenation and other side reactions. For the effect of pressure, the crude biodiesel and FAMEs yield increased with increasing pressure. Above 12 MPa, no improvement of both yields was observed. The optimum pressure was thus fixed at 12.7 MPa in this experiment. For the effect of reaction time, it can be seen that the conversion was increased in the reaction time ranges between 5 and 30 min, and thereafter reduced as a representative of the equilibrium conversion. The excess reaction time did not promote the conversion but favors the reverse reaction of transesterification which resulted in a reduction in the ester yield. The optimal FAMEs yield was found to be 97.9% in 30 min. For the effect of seeds-to-methanol ratio, the maximum crude biodiesel and FAMEs yields were obtained at a 1:40 of seeds-to-methanol ratio. It can be seen that an increment in seed-to-methanol ratio can enhance biodiesel yield due to higher contact area between methanol and triglycerides. However, when the ratio is beyond 40, the yield of biodiesel begins to decrease substantially.

The merit of this method is that this new process just requires a single process, where the normal oil extraction process can be avoided. In addition, because of non-catalytic process, the purification of products after transesterification reaction is much simple, compared to the common method. Therefore, this new process can offer an alternative way to convert the fruits directly to methyl esters by a simpler-shorter production process.

5. Acknowledgment

The authors ackonowledge the scholarship fund provided by Ministry of Science, Technology and Innovative under the National Science Fellowship (NSF), Universiti Teknologi MARA and Fundamental Research Grant Scheme (FRGS grant no: 600-RMI/ST/FRGS 5/3/Fst (2/2009)) for their financial support.

6. References

Ayuso, L. E. G. & Castro M. D. L. (1999). A multivariate study of the performance of a microwave-assissted Soxhlet extractor for olive seeds. *Analytica Chimica Acta*, Vol. 382(3), pp. 309-316.

Berchmans, H. J. & Hirata, S. (2008). Biodiesel production from crude *Jatropha curcas* L. seed oil with a high content of free fatty acids. *Bioresource Technology*, Vol. 99, pp. 1716–1721.

Bunyakiat, K.; Makmee, S.; Sawangkeaw, R. & Ngamprasertsith. (2006). Continous production of biodiesel via transesterification from vegetable oils in supercritical methanol. *Energy & Fuel*, Vol. 20, pp. 812-817.

Cao, W.; Han, H. & Zhang, J. (2005). Preparation of biodiesel from soybean oil using supercritical methanol and co-solvent. *Fuel*, Vol. 84(4), pp. 347-351.

Cao, X. & Ito, Y. (2003). Supercritical fluid extraction of grape seed oil and subsequent separation of free fatty acids by high-speed counter-current chromatography. *Journal of Chromatography A*, Vol. 1021, pp. 117–124.

Carrapiso, A. I. and García, C. (2000). Development in lipid analysis: some new extraction techniques and in situ transesterification. *Lipids*, Vol. 35, pp. 1167–1177.

Castro, M. D. L. & Ayuso, L. E. G. (1998). Soxhlet extraction of solid materials: Soxhlet extraction of solid materials: anoutdated technique with a promising innovative future. *Analytica Chimica Acta*, Vol. 369(1-2), pp. 1-10.

Chen, W. H.; Chen, C. H.; Chang, C. M. J.; Chiu, Y. H. & Hsiang, D. (2009). Supercritical carbon dioxide extraction of triglycerides from *Jatropha curcas* L. seeds. *Journal of Supercritical Fluids*, Vol. 51, pp. 174-180.

Chew, C. S. (2009). *Jatropha curcas*. L.: Development of a new oil crop for biofuel (Summary). New and Renewable Energy Group, Strategy and Industry Research Unit, The Institute of Energy Economics, Japan (IEEJ).

Clark, S. J.; Wagner, L.; Schrock, M. D. & Pinnaar, P. G. (1984). Methyl and ethyl esters as renewable fuels for diesel engines. *Journal of American Oil Chemists' Society*, Vol. 61, pp. 1632-1638.

Darcia, J. L. L. & Castro, M. D. L. (2004). Ultrasound-assisted Soxhlet extraction: An expeditive approach for solid sample treatment – Application to the extraction of total fat from oleaginous seeds. *Journal of Chromatography A*, Vol. 1034 (1-2), pp. 237-242.

De, B. K.; Bhattacharyya, D. K. & Bandhu, C. (1999). Enzymatic synthesis of fatty alcohol esters by alcoholysis. *Journal of American Oil Chemists' Society*, Vol. 76, pp. 451-453.

Demirbas, A. (2003). Biodiesel fuels from vegetable oils via catalytic and non-catalytic supercritical alcohol transesterifications and other methods: a survey. *Energy Convers Mgmt, Vol.* 44, pp. 2093-109.

Demirbas, A. (2005). Biodiesel production from vegetable oils via catalytic and non catalytic supercritical methanol transesterification methods. *Progress in Energy and Combustion Science, Vol.* 31, pp. 466-487.

Demirbas, A. (2007). Biodiesel from sunflower oil in supercritical methanol with calcium oxide. *Energy Conversion and Management*, Vol. 48(3), pp. 937-941.

Demirbas, A. (2008). Studies On cottonseed oil biodiesel prepared in non-catalytic SCF conditions. *Bioresource Technology*, Vol. 99(5), pp. 1125-1130.

Diasakou, M.; Louloudi, A. & Papayannakos, N. (1998). Kinetics of the non-catalytic transesterification of soybean oil. *Fuel*, Vol. 77(12), pp. 1297-1302.

Eevera, T.; Rajendran, K. & Saradha, S. (2009). Biodiesel production process optimization and characterization to assess the suitability of the product for varied environmental conditions. *Renewable Energy*, Vol. 34(3), pp. 762-765.

Encinar, J. M.; Gonzalez, J. F. & Reinares, A. R. (2005). Biodiesel from Used Frying Oil. Variables Affecting the Yields and Characteristics of the Biodiesel. *Industrial & Engineering Chemistry Research*, Vol. 44(15), pp. 5491-5499.

Fernando, S.; Karra, P.; Hernandez, R. & Jha, S. K. (2007). Effect of incompletely converted soybean oil on biodiesel quality. *Energy*, Vol. 32, pp. 844-851.

Freedman, B.; Pyrde, E. H. & Mounts, T. L. (1984). Variables affecting the yield of fatty esters from transesterified vegetable oils. *Journal of American Oil Chemists' Society*, Vol. 61(10), pp. 1638-1643.

Fukuda, H.; Kondo, A. & Noda, H. (2001). Biodiesel fuel production by transesterification of oils. *Journal of Bioscience and Bioengineering*, Vol. 92(5), pp. 405-416.

Gaur, R.; Sharma, A.; Khare, S. K. & Gupta, M. N. (2007). A novel process for extraction of edible oils: Enzyme assisted three phase portioning (EATPP). *Bioresource Technology*, Vol. 98(3), pp. 696-699.

Ghadge, S. V. & Raheman, H. (2005). Biodiesel production from mahua (Madhuca indica) oil having high free fatty acids. *Biomass and Bioenergy*, Vol. 28, pp. 601–605.

Han, H.; Cao, W. & Zhang, J. (2005). Preparation of biodiesel from soybean oil using supercritical methanol and CO_2 as co-solvent. *Process Biochemistry, Vol.* 40, pp. 3148-3151.

Harrington, K. J. & Evans, C. D. A. (1985). Transesterification in situ of sunflower seed oil. *Industrial & Engineering Chemistry Product Research & Development*, Vol. 62, pp. 314–318.

Hass, M. J.; Scott, K. M.; Marmer, W. N. & Foglia, T. A. (2004). In situ alkaline transesterification: an effective method for the production of fatty acid esters from vegetable oils, *Journal of American Oil Chemists' Society*, Vol. 81, pp. 83–89.

Hawash, S.; Kamal, N.; Zaher, F.; Kenawi, O. & El Dewani, G. (2009). Biodiesel fuel from *Jatropha* oil via non-catalytic supercritical methanol transesterification. *Fuel*, Vol. 88(3), pp. 579-582.

He, H.; Wang, T. & Zhu, S. (2007). Continuous production of biodisel fuel from vegetable oil using supercritical methanol process. *Fuel, Vol.* 86, pp. 442-447.

Heller, J. (1996). *Physic nut. Jatropha curcas L. Promoting the conservation and use of underutilized and neglected crops.* Institute of Plant Genetics and Crop Plant Research, Gatersleben/ International Plant Genetic Resources Institute, Rome.

Herrera, J. M.; Siddhuraju, P.; Francis, G., Ortiz.; G. D. V. & Becker, K. (2006). Chemical composition, toxic/antimetabolic constituents, and effects of different treatments on their levels, in four provenances of *Jatropha curcas* L. from Mexico. *Food Chemistry*, Vol. 96(1), pp. 80–89.

Ikwuagwu, O. E.; Ononogbu, I. C. & Njoku, O. U. (2000). Production of biodiesel using rubber [*Hevea brasiliensis (Kunth. Muell.)*] seed oil. *Industrial Crops and Products*, Vol. 12(1), pp. 57-62.

Jiang, L.; Hua, D.; Wang, Z. & Xu, S. (2009). Aqueous enzymatic extraction of peanut oil and protein hydrolysates. *Food and Bioproducts Processing*, Vol. 88 (2-3), pp. 2010.

Kansedo, J.; Lee, K. T. & Bhatia, S. (2009). *Cerbera odollam* (sea mango) oil as promising non-edible feedstock for biodiesel production. *Fuel*, Vol. 88(6), pp. 1148-1150.

Kildiran, G.; Yucel, S. O. & Turkay, S. (1996). In-situ alcoholysis of soybean oil, *Journal of American Oil Chemists' Society*, Vol. 73, pp. 225–228.

King, J. W.; Mohamed, A.; Taylor, S. L.; Mebrahtu, T. & Paul, C. (2001). Supercritical fluid extraction of *Vernonia galamensis* seeds. *Industrial Crops and Products*, Vol. 14, pp. 241–249.

Kusdiana, D. & Saka, S. (2001). Kinetics of transesterification in rapeseed oil to biodiesel fuel as treated in Supercrititcal methanol. *Fuel*, Vol. 80, pp. 693-698.

Kusdiana, D. & Saka, S. (2004). Effects of water on biodiesel fuel production by supercritical methanol treatment. *Bioresource Technology*, Vol. 91, pp. 289–295.

Leung, D. Y. C.; Koo, B. C. P. & Guo, Y. (2006). Degradation of biodiesel under different storage conditions. *Bioresource Technology*, Vol. 97(2), pp. 250-256.

Lu, H.; Liu, Y.; Yang, Y.; Chen, M. & Liang, B. (2009). Production of biodiesel from *Jatropha curcas* L. oil. *Computers and Chemical Engineering*, Vol. 33 (5), pp. 1091-1096.

Ma, F.; Clements, L. D. & Hanna, M. A. (1998). The effects of catalysts, free fatty acids and water on transesterification of beef tallow. *Transactions of the American Society of Agricultural Engineers*, Vol. 41, pp. 1261-1264.

Machmudah, S.; Kawahito, Y.; Sasaki, M. & Goto, M. (2007). Supercritical CO_2 extraction of rosehip seed oil: Fatty acids composition and process optimization. *Journal of Supercritical Fluids*, Vol. 41(3), pp. 421-428.

Machmudah, S.; Kondo, M.; Sasaki, M.; Goto, M.; Munemasa, J. & Yamagata, M. (2008). Pressure effect in supercritical CO_2 extraction of plant seeds. *Journal of Supercritical Fluids*, Vol. 44, pp. 301-307.

Madras, G.; Kolluru, C. & Kumar, R. (2004). Synthesis of biodiesel in supercritical fluids. *Fuel, Vol.* 83, pp. 2029-2033.

Marinkovic, S. S. &Tomasevic, A. (1998). Transesterification of sunflower oil in situ. *Fuel*, Vol. 77(12), pp. 1389-1391.

Meher L. C.; Kulkarni M. G.; Dalai A. K. & Naik, S. N. (2006). Transesterification of karanja (Pongamia pinnata) oil by solid basic catalysts. *European Journal of Lipid Science & Technology*, Vol. 108, pp. 389-397.

Murugesan, A.; Umarani, C.; Chinnusamy, T. R.; Krishnan, M.; Subramanian, R. & Neduzchezhain, N. (2009). Production and analysis of bio-diesel from non-edible oils — A review. *Renewable and Sustainable Energy Reviews*, Vol. 13, pp. 825-834.

Naik, M.; Meher, L. C.; Naik, S. N. & Das, L. M. (2008). Production of biodiesel from high free fatty acid Karanja (*Pongamia pinnata*) oil. *Biomass and Bioenergy*, Vol. 32(4), pp. 354-357.

Papamichail, I.; Louli, V. & Magoulas, K. (2000). Supercritical fluid extraction of celery seed oil. *Journal of Supercritical Fluids*, Vol. 18, pp. 213-226.

Petchmala, A.; Yujaroen, D.; Shotipruk, A.; Goto, M. & Sasaki, M. (2008). Production methyl esters from palm fatty acids in supercritical methanol. Chiang Mai Journal *of Science*, Vol. 35(1), pp. 23-28.

Pradhan, R. C.; Naik, S. N.; Bhatnagar, N. & Swain, S. K. (2008). Moisture-dependent physical properties of Karanja (*Pongamia pinnata*) kernel. *Industrial Crops and Products*, Vol. 28(2), pp. 155-161.

Qian, J.; Wang, F.; Liu, S. & Yun, Z. (2008). In situ alkaline transesterification of cottonseed oil for production of biodiesel and nontoxic cottonseed meal. *Bioresource Technology*, Vol. 99(18), pp. 9009-9012.

Raheman, H. & Ghadge, S. V. (2007). Performance of compression ignition engine with mahua (*Madhuca indica*) biodiesel. *Fuel*, Vol. 86(16), pp. 2568-2573.

Ramadhas, A. S.; Jayaraj, S. & Muraleedharan, C. (2005). Biodiesel production from high FFA rubber seed oil. *Fuel*, Vol. 84(4), pp. 335-340.

Rashid, U.; Anwar, F. & Knothe, G. (2009). Evaluation of biodiesel obtained from cottonseed oil. *Fuel Processing Technology*, Vol. 90(9), pp. 1157-1163.

Rathore, V. & Madras, G. (2007). Synthesis of biodiesel from edible and non-edible oils in supercritical alcohols and enzymatic synthesis in supercritical carbon dioxide. *Fuel,* Vol. 86, pp. 2650-2659.

Rosenthal, A.; Pyle, D. L. & Niranjan, K. (1996). Aqueous and enzymatic processes edible oil extraction for edible oil extraction. *Enzyme and Microbial Technology*, Vol. 19, pp. 402-420.

Sahoo, P. K. & Das, L. M. (2009). Process optimization for biodiesel production from *Jatropha*, karanja and *Polanga* oils. *Fuel*, Vol. 88(9), pp. 1588-1594.

Sahoo, P. K.; Das, L. M.; Babu, M. K. G. & Naik, S. N. (2007). Biodiesel development from high acid value polanga seed oil and performance evaluation in a CI engine. *Fuel*, Vol. 86(3), pp. 448-454.

Saka, S. & Kusdiana, D. (2001). Biodiesel fuel from rapeseed oil as prepared in supercritical methanol. *Fuel*, Vol. 80, pp. 225-231.

Selmi, B. & Thomas, D. (1998). Immobilized lipase-catalyzed ethanolysis of sunflower oil in a solvent-free medium. *Journal of American Oil Chemists' Society*, Vol. 75, pp. 691-695.

Sharma, A. & Gupta, M. N. (2006). Ultrasonic pre-irradiation effect upon aqueous enzymatic oil extraction from almond and apricot seeds. *Ultrasonics Sonochemistry*, Vol. 13(6), pp. 529-534.

Sinha, S.; Agarwal, A. K. & Garg, S. (2008). Biodiesel development from rice bran oil: Transesterification process optimization and fuel characterization. *Energy Conversion and Management*, Vol. 49(5), pp. 1248-1257.

Tan, K. T.; Lee,K. T. & Mohamed, A. R. (2009). Production of FAME by palm oil transesterification via supercritical methanol technology. *Biomass and Bioenergy*, Vol. 33(8), pp. 1096-1099.

Tiwari, A. K.; Kumar, A. & Raheman, H. (2007). Biodiesel production from *Jatropha* oil (*Jatropha Curcas*) with high free fatty acids: an optimized process", *Biomass and Bioenergy*, Vol. 31, pp. 569-575.

Usta, N. (2005). An experimental study on performance and exhaust emissions of a diesel engine fuelled with tobacco seed oil methyl ester. *Energy Conversions and management*, Vol. 46(15-16), pp. 2373-2386.

Varma, M. N. & Madras, G. (2007). Synthesis of biodiesel from castor oil and linseed oil in supercritical fluids, *Industrial & Engineering Chemistry Research*, Vol. 46, pp. 1–6.

Veljkovic, V. B.; Lakicevic, S. H.; Stamenkowc, O. S.,;Todorovic, Z. B. & Lazic, M. L. (2006). Biodiesel production from tobacco (*Nicotianna tabacum* L.) seed oil with a high content of free fatty acids. *Fuel*, Vol. 85(17-18), pp. 2671-2675.

Vyas, A. P.; Subrahmanyam, N. & Patel, A. (2009). Production of biodiesel through transesterification of *Jatropha* oil using KNO_3/Al_2O_3 solid catalyst. *Fuel*, Vol. 88(4), pp. 625-628.

Wang, Y.; Ou, S.; Liu, P.; Xue, F. & Tang, S. (2006). Comparison of two different processes to synthesize biodiesel by waste cooking oil. *Journal of Molecular Catalysis A: Chemical*, Vol. 252(1-2), pp. 107-112.

Warabi, Y.; Kusdiana, D. & Saka, S. (2004). Reactivity of triglycerides and fatty acids of rapeseed oil in supercritical alcohols. *Bioresource Technology*,Vol. 9, pp. 283-287.

Xin, J.; Imahara, H. & Saka, S. (2008). Oxidation stability of biodiesel fuel as prepared by supercritical methanol. *Fuel*, Vol. 87(10-11), pp. 1807-1813.

Zhang, Y.; Dube, M. A.; McLean, D. D. & Kates, M. (2003). Biodiesel production from waste cooking oil: 1. Process design and technological assessment. *Bioresource Technology*, Vol. 89(1), pp. 1-16.

4

Transesterification in Supercritical Conditions

Somkiat Ngamprasertsith and Ruengwit Sawangkeaw
Fuels Research Center, Department of Chemical Technology,
Faculty of Science, Chulalongkorn University,
Center for Petroleum, Petrochemicals and Advance Materials,
Chulalongkorn University,
Thailand

1. Introduction

The transesterification or biodiesel production under supercritical conditions (supercritical transesterification) is a catalyst-free chemical reaction between triglycerides, the major component in vegetable oils and/or animal fats, and low molecular weight alcohols, such as methanol and ethanol, at a temperature and pressure over the critical point of the mixture (see Section 1.1). The overall transesterification reaction is shown in Fig. 1.

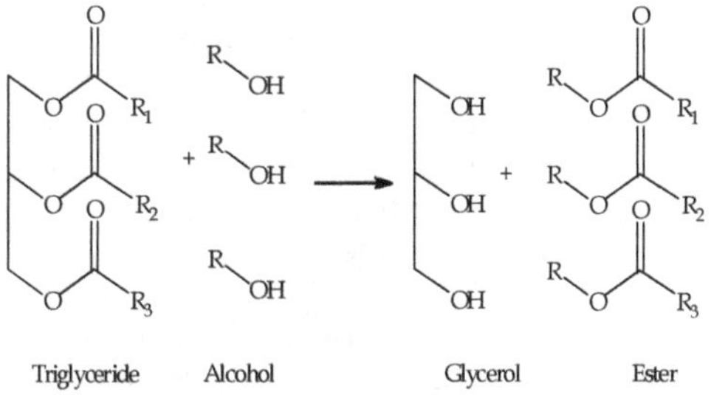

Triglyceride   Alcohol   Glycerol   Ester

Fig. 1. The overall transesterification reaction (R is a small alkyl group, R_1, R_2 and R_3 are a fatty acid chain)

The reaction mechanism for supercritical transesterification has been proposed to be somewhat alike the acidic-catalyzed reaction as described in Section 1.2.

Since the actual feedstocks are not composed solely of triglycerides, especially the low-grade feedstocks, but are also contaminated with water and free fatty acids, some side reactions also take place under supercritical conditions (see Section 1.3). For example, the esterification of free fatty acids with alcohols increases the fatty acid alkyl ester content in the biodiesel product, while the thermal cracking of unsaturated fatty acids decreases the esters content.

The earlier research on supercritical transesterification mostly employed methanol as the reacting medium and reacting alcohol at the same time due to the fact that it has the lowest critical point and the highest activity (Warabi et al., 2004). Ethanol is also an interesting candidate because it can be industrially produced from renewable sources in many countries nowadays. However, other supercritical mediums, such as methyl acetate (Saka & Isayama, 2009) and dimethyl carbonate (Ilham & Saka, 2009; Tan et al., 2010b), have also been used to produce biodiesel, but these are not be described in this chapter since the chemical reaction involved is not a transesterification reaction. The reaction parameters and optimal conditions for supercritical transesterification are summarized in Section 1.4.

1.1 The definition of supercritical transesterification

A pure substance ordinarily exists in a solid, liquid or gaseous state, depending on the temperature and pressure. For example, methanol is in a liquid state at ambient temperature (and pressure) and changes to a gaseous state above its boiling point. A gaseous substance can be compressed to a liquid state when a pressure above the boiling point is applied. Until the critical temperature is reached, a gaseous substance cannot be compressed to the liquid state. In the same manner, a compressed liquid substance cannot be heated to a gaseous state at its critical pressure.

Above its inherent critical temperature and pressure, the substance becomes a supercritical fluid, which is a non-condensable dense fluid. In the supercritical state, the density is generally in a range between 20 – 50% of that in the liquid state and the viscosity is close to that in the gaseous state. In other words, the molecules in the supercritical fluid have high kinetic energy like a gas and high density like a liquid. Therefore, the chemical reactivity can be enhanced in this state.

The critical point of any transesterification reaction mixture is mostly calculated by the critical properties of the alcohols and the vegetable oils and/or animal fats. However, the critical properties of vegetable oils and/or animal fats cannot be experimentally measured because they thermally decompose before the critical point is reached. In addition, the molecular structure of vegetable oils and/or animal fats is impossible to know because the exact distribution of the fatty acids chain in triglycerides mixture is unknown.

Therefore to estimate the critical properties of vegetable oils and/or animal fats, their molecular is assumed to be a simple triglyceride (tripalmitin, triolein, etc.) or pseudo-triglycerides (Espinosa et al., 2002), with the proportion of such different simple triglycerides reflecting the actual overall fatty acid composition in the feedstock. The type of simple triglycerides or the pseudo-triglycerides are thus defined by their actual fatty acid profile in the vegetable oils and/or animal fats. For instance, soybean oil has linoleic acid as the major fatty acid, and so it is usually assumed to be trilinolein. Next, the critical properties of the simple triglycerides or the pseudo-triglycerides are estimated by the Fedor and Lydersen group contribution method (Poling et al., 2001), or similar. After the critical properties of each of the triglycerides are estimated, the critical point of the mixture can be estimated by mixing rules, such as the Lorentz-Berthelot-type (Bunyakiat et al., 2006) and the group-contribution with associated mixing rules (Hegel et al., 2008).

For example, the critical temperature and pressure of soybean oil-methanol and palm kernel oil-methanol mixtures are illustrated in Figs. 2 and 3, respectively.

From Figs. 3 and 4, it is clear that the critical point of the reaction mixture depends on the alcohol to oil molar ratio, so the selected alcohol to oil molar ratio will reflect the operating temperature and pressure, as described in Section 2.1. For a high transesterification conversion (triglyceride to alkyl ester) at a constant methanol to oil molar ratio, the operating temperature and pressure have to be approximately 1.5- to 2.0-fold over the critical point of the reaction mixture. For example, the optimal conditions at a methanol to oil molar ratio of 42:1 is 350 °C and 20 MPa, respectively. Therefore, the definition of supercritical conditions is the temperature and pressure above the critical point of the reaction mixture, which is calculated from the critical properties of the vegetable oils and/or animal fats and the alcohols.

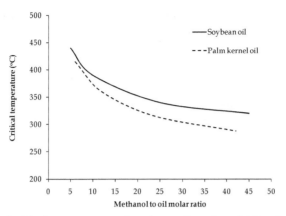

Fig. 2. The estimated critical temperature of soybean oil-methanol (Hegel et al., 2008) and palm kernel oil-methanol (Bunyakiat et al., 2006) mixtures.

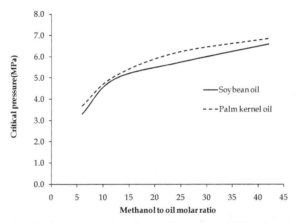

Fig. 3. The estimated critical pressure of soybean oil-methanol (Hegel et al., 2008) and palm kernel oil-methanol (Bunyakiat et al., 2006) mixtures.

1.2 The reaction mechanism of supercritical transesterification

Below the critical point of mixture, transesterification can take place in the presence of acidic or basic catalysts. Thus, the reaction mechanisms of transesterification are divided into acid and base catalyzed paths, as summarized elsewhere (Meher et al., 2006). The reaction mechanism of supercritical transesterification is somewhat similar to the acid catalyzed path in that the hydrogen bond of the alcohol is weakened at high temperatures (Hoffmann & Conradi, 1998). However, whilst the acid-catalyzed transesterification reaction is much slower than the base-catalyzed one at ambient conditions, the supercritical transesterification is much faster and achieves complete conversion of triglycerides to esters rapidly because the chemical kinetics are dramatically accelerated under supercritical conditions.

The reaction mechanism of supercritical transesterification, as shown in Fig. 4, was proposed by the analogues between the hydrolysis of esters in supercritical water (Krammer

& Vogel, 2000) and the transesterification of triglycerides in supercritical methanol (Kusdiana & Saka, 2004b). It is assumed that the alcohol molecule (in this case methanol) directly attacks the carbonyl carbon of the triglyceride because the hydrogen bond energy is lowered; which would allow the alcohol to be a free monomer. In the case of methanol, the transesterification is completed via transfer of a methoxide moiety, whereby fatty acid methyl esters and diglycerides are formed. Consequently, the diglyceride reacts with other methanol molecules in a similar way to form the methyl ester and monoglyceride, the later of which is further converted to methyl ester and glycerol in the last step. The same process is applicable to other primary alkyl alcohols, such as ethanol.

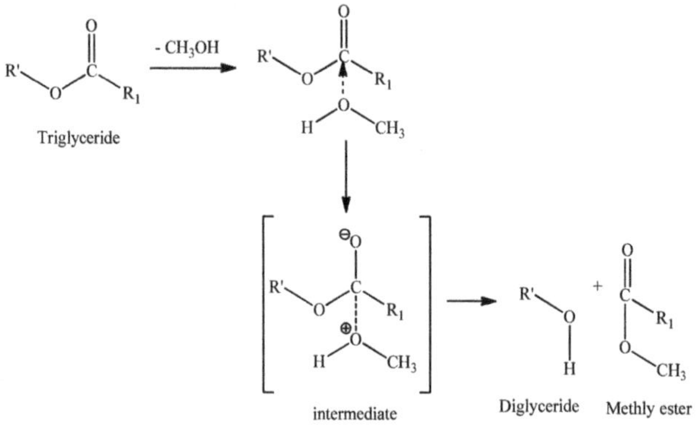

Fig. 4. The proposed reaction mechanism of transesterification in supercritical methanol (R' is a diglyceride group and R_1 is a fatty acid chain).

1.3 The side reactions in the supercritical conditions

Firstly, the hydrolysis reaction of alkyl esters and triglyceride can take place at over 210 °C (Khuwijitjaru et al., 2004) and over 300 °C (W. King et al., 1999), respectively, in present of water and pressure above 20 MPa. The overall hydrolysis reaction of triglycerides and alkyl esters is shown in Figs. 5 and 6, respectively, with the principal products of the hydrolysis reaction being the respective fatty acids which are subsequently converted to alkyl esters under the supercritical condition by the esterification, as illustrated in Fig. 7.

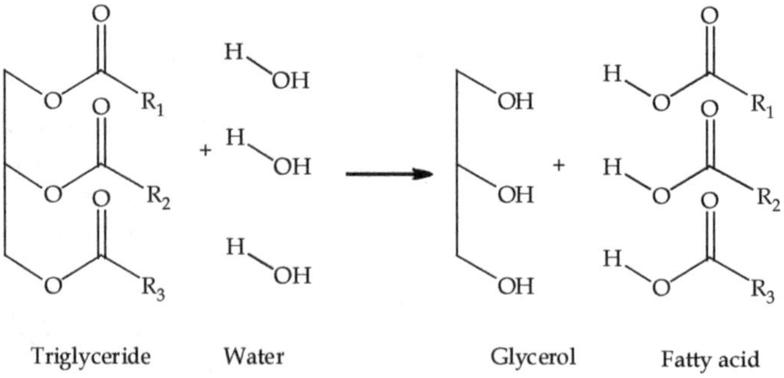

Fig. 5. The overall hydrolysis reaction of triglycerides under supercritical conditions (R is an alkyl group and R_1, R_2 and R_3 are fatty acid chains)

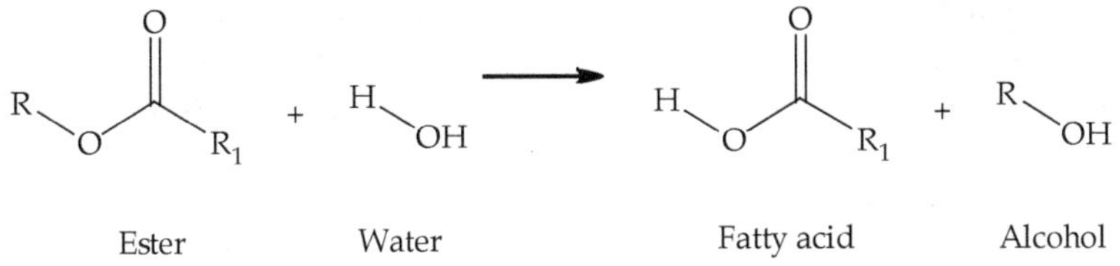

Ester   Water   Fatty acid   Alcohol

Fig. 6. The hydrolysis reaction of alkyl esters under supercritical conditions (R is an alkyl group and R_1 is a fatty acid chain)

Fatty acid   Alcohol   Ester   Water

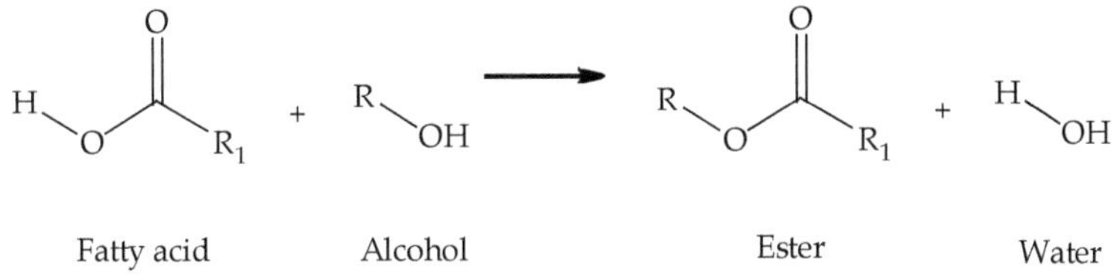

Fig. 7. The esterification reaction under supercritical conditions (R is an alkyl group and R_1 is a fatty acid chain)

In actuality, the hydrolysis reaction or the presence of water and free fatty acids do not affect the final alkyl ester content obtained for supercritical transesterification (Kusdiana & Saka, 2004b), because the alcohols have a much higher reactivity than water at the optimal point of supercritical transesterification. For example, the chemical rate constant for the transesterification of rapeseed oil is approximately 7-fold higher than that for the hydrolysis of soybean oil (Khuwijitjaru et al., 2004; Kusdiana & Saka, 2001).

Secondly, the thermal cracking of unsaturated fatty acids, especially the polyunsaturated fatty acids, occurs at temperatures over 300 °C and reaction times over 15 min (Quesada-Medina & Olivares-Carrillo, 2011). An example of the thermal cracking of a palmitic, oleic and linoleic acid based triglyceride is illustrated in Fig. 8.

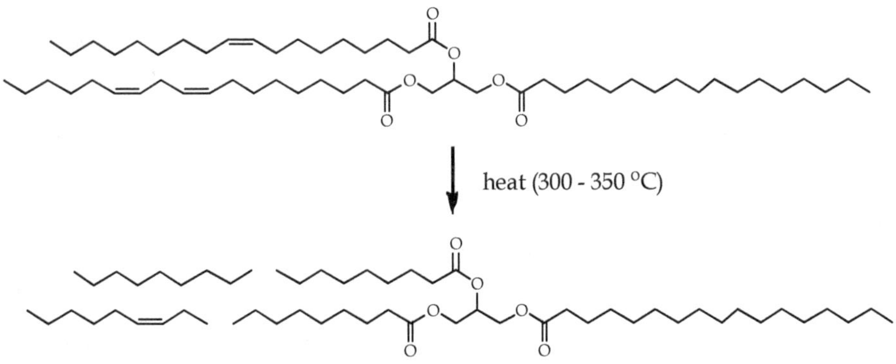

heat (300 - 350 °C)

Fig. 8. The thermal cracking reaction of a triglyceride under supercritical conditions at a temperature range of 300 - 350 °C and a reaction time of over 15 min.

The triglyceride product from the thermal cracking reaction can be transesterified afterwards under supercritical conditions to alkyl esters. However, these alkyl esters are not the fatty acid alkyl esters of the common fatty acids in vegetable oils and/or animal fats. Therefore, the thermal cracking reaction reduces the acceptable primary alkyl ester content, as quantified and defined by the international standard for biodiesel (EN14103), and this is especially the case for the oils with a high polyunsaturated fatty acid content, such as soybean and sunflower oil.

In addition, triglycerides are decomposed to fatty acids and some gaseous products within the temperature range of 350 - 450 °C, as shown in Fig. 9 (Lima et al., 2004; Marulanda et al., 2009). In the same manner, with thermal cracking at 300 - 350 °C, the fatty acids product can be esterified under supercritical conditions, but the alkyl ester content is also decreased. However, the small hydrocarbon molecules of the thermal cracking products could improve some fuel properties of biodiesel, such as viscosity, density and cold flow properties.

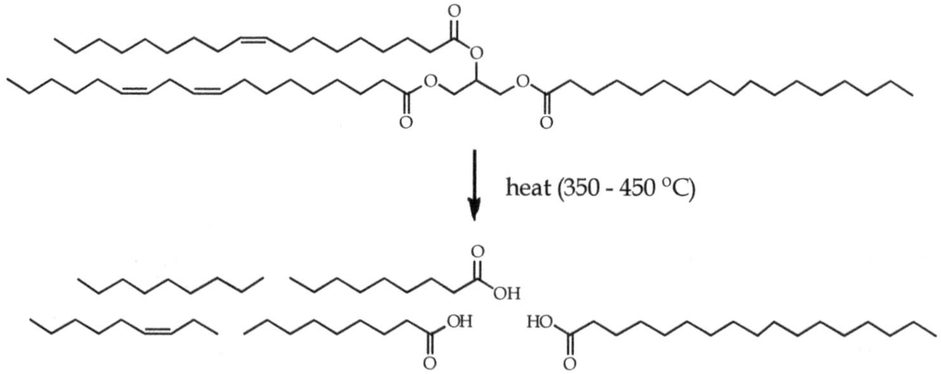

Fig. 9. The thermal cracking reaction of triglycerides under supercritical conditions at temperature range of 350 - 450 °C

1.4 The original reaction parameters and optimal conditions in the early scientific articles

The reaction parameters that were typically investigated in supercritical transesterification reactions are the temperature, pressure, alcohol to oil molar ratio and reaction time in batch and continuous reactors, and are summarized in Table 1.

The extent of the reaction was reported in terms of the % alkyl ester content and the % conversion of triglycerides. The % alkyl esters content refers to the alkyl esters of the common fatty acids in the vegetable or animal oils/fats that can be identified by different analytical techniques, while the % triglycerides conversion implies the remaining triglyceride reactant that is converted to fuels. Note that the % alkyl esters content refers to the specified esters, which must not be less than 96.5%, in the International standard (EN14214) for biodiesel fuel. It should also be noted that a high alkyl esters content infers a high triglyceride conversion, but, in contrast, a high triglyceride conversion does not have to infer a high alkyl esters content because the triglycerides could have been converted by the side reactions to other products.

According to Table 1, the original optimal conditions, defined as yielding the highest extent of reaction as over 90% conversion or over 96% alkyl esters content, were within 300 – 350 °C, 20 – 35 MPa, an alcohol to oil molar ratio of 40:1 – 42:1 and a reaction time of

5 – 30 min, for both methanol and ethanol. These parameters are referred to as the original supercritical transesterification parameters, and have been employed to study the effects of each parameter, the chemical kinetics, the phase behavior and the economical feasibility of the process. Since the original parameters are elevated conditions, innovative techniques are proposed to reduce these original parameters.

2. Process overview

2.1 The effects of reaction parameters on the % conversion in supercritical transesterification

Among the general operating parameters mentioned previously (temperature, pressure, alcohol to oil molar ratio and reaction time), the reaction temperature is the most decisive parameter for indicating the extent of the reaction. This is as a result of the accelerated chemical kinetics and changes to the alcohol's properties. For example, the rate constant of supercritical transesterification is dramatically enhanced some 7-fold as the temperature is increased from 210 to 280 °C at 28.0 MPa and a 42:1 methanol to oil molar ratio (He et al., 2007a), whilst the degree of hydrogen bonding also suddenly drops as the temperature is increased from 200 to 300 °C at 30.0 MPa (Hoffmann & Conradi, 1998). However, where a maximum alkyl ester content is required, that is for biodiesel production, the higher operating temperatures cause a negative effect on the proportion of alkyl esters obtained in the product due to the thermal cracking reaction. Indeed, the thermal cracking is the chemical limitation of supercritical transesterification, and this is discussed in Section 2.4.2.

Oil type	Alcohol	T (°C)	P (MPa)	Alcohol: oil (mol: mol)	Reaction time (min)	Reactor (size / type)[a]	Extent of reaction[b]	Ref
Coconut & Palm kernel	Methanol	350	19	42:1	7 - 15	251-mL / CT	95% MEC	(Bunyakiat et al., 2006)
Hazelnut kernel & Cottonseed	Methanol	350	NR	41:1	5	100-mL / Batch	95% MEC	(Demirbas, 2002)
Palm and Groundnut	Methanol	400	20	50:1	30	11-mL / Batch	95% Con.	(Rathore & Madras, 2007)
Palm kernel	Methanol	350	20	42:1	30	250-mL / Batch	95% MEC	(Sawangkeaw et al., 2007)
Rapeseed	Methanol	350	45	42:1	4	5-mL / Batch	98% MEC	(Saka & Kusdiana, 2001)
Rapeseed	Methanol	350	20	42:1	30	200-mL / CT	87% MEC	(Minami & Saka, 2006)
Soybean	Methanol	350	20	42:1	30	250-mL / Batch	95% MEC	(Yin et al., 2008a)
Sunflower	Methanol & Ethanol	400	20	40:1	30	8-mL / Batch	97% Con.	(Madras et al., 2004)
Castor	Ethanol	300	20	40:1	NR	42-mL / CT	75% EEC	(Vieitez et al., 2011)
Soybean	Ethanol	350	20	40:1	15	42-mL / CT	80% Con.	(Silva et al., 2007)
Sunflower	Ethanol	280	NR	40:1	5	100-mL / Batch	80% EEC	(Balat, 2008)

[a] CT = Continuous reaction in a tubular vesicle.
[b] Reaction extents are expressed as the % triglyceride conversion (Con.), % methyl esters content (MEC) or the % ethyl esters content (EEC).
NR = not reported

Table 1. The original reaction parameters and optimal conditions of supercritical transesterification for various oil types and alcohols.

The reaction pressure also has a significant effect on the efficiency of the supercritical transesterification reaction below 20.0 MPa, but the effects tend to be negligible above 25.0 MPa (He et al., 2007a; He et al., 2007b), due to the fact that increasing the reaction pressure simultaneously increases both the density of the reaction mixture (Velez et al., 2010) and the degree of hydrogen bonding (Hoffmann & Conradi, 1998) at an otherwise constant temperature and alcohol to oil molar ratio. The transesterification conversion is enhanced with an increased reaction mixture density, due to the resulting increased volumetric concentration of alcohols and the residence time in a tubular reactor, which is commonly used to investigate the effect of pressure. On the other hand, the increasing degree of hydrogen bonding or alcohol cluster size weakens the nucleophilic strength of supercritical alcohols and so the reactivity of supercritical alcohol is reduced with increasing pressure. Thus, the desirable pressure for supercritical transesterification is in the range of 20.0 – 35.0 MPa.

From Table 1, the original alcohol to oil molar ratio for supercritical transesterification is in range of 40:1 – 42:1. The alcohol to oil molar ratio affects the supercritical transesterification efficiency strongly below 24:1, but its effect is then reduced with increasing alcohol to oil molar rations to plateau at over a 50:1 alcohol to oil molar ratio for methanol (He et al., 2007a) or 70:1 for ethanol (Silva et al., 2007) at 330 °C and 20.0 MP. This is likely to be due to the fact that the operating temperature and pressure are much higher than the critical point of the reaction mixture, and are still located in the supercritical region.

As mentioned in Section 1.1, the critical point of the reaction mixture decreases with increasing alcohol to oil molar ratios. Thus, the optimal reaction temperature and pressure at a high alcohol to oil molar ratio is always milder than that at a low molar ratio. Nonetheless, the large amount of alcohol not only increases the required reactor volume but importantly it consumes a large amount of energy to heat the reactant and also to subsequently recover the excess alcohol. The energy for recycling excess alcohol might be minimized by a low temperature separation process, such as the use of a medium-pressure flash drum (Diaz et al., 2009), whereas the additional energy for heating the excess reactant alcohol cannot be avoided. Therefore, the use of assisting techniques, as described in Sections 3.2 – 3.4, have been introduced to decrease the alcohol to oil molar ratio whilst maintaining the transesterification conversion efficiency (and so the fatty acid alkyl ester content).

The effect of the reaction time on the transesterification conversion follows the general rate law. For example, the alkyl ester content increases gradually with reaction time and then remains constant after the optimal point (maximum conversion to alkyl esters) is reached. The optimal reaction time for supercritical transesterification at around 300 – 350 °C varied between 4 to 30 minutes, depending upon the reactor size and type. Since the effect of residence time is directly related to the chemical kinetics of transesterification, the optimal reaction time at low temperatures is longer than that at high temperatures.

2.2 The chemical kinetics and phase behavior in supercritical transesterification
2.2.1 The chemical kinetics of supercritical transesterification
The chemical kinetics of supercritical transesterification is divided into three regions, that of the slow (<280 °C), transition (280 - 330 °C) and fast (>330 °C) regions, and usually follows the first-order rate law with respect to the triglyceride concentration alone (He et al., 2007a; Kusdiana & Saka, 2001; Minami & Saka, 2006). Here, the reaction mechanism is merged into one overall step and the concentrations of all intermediates (mono- and diglycerides) are ignored. However, the first-order kinetic model is only suitable for a high alcohol to oil

molar ratio, due to the insignificant changes in the alcohol concentration, but this increasingly becomes untrue as the alcohol to oil molar ratio decreases.

For the first-order model, the rate constants for each vegetable oil have a different temperature sensitivity, as noticed by the slope of Arrhenius' plot (Sawangkeaw et al., 2010). For example, the rate constants of rapeseed and soybean oil depend more strongly on the temperature than that for sunflower, palm and groundnut oils. The rate constants of saturated triglycerides were found to be faster than unsaturated triglycerides and slow down with increasing levels of double bonds in the triglyceride molecule (Rathore & Madras, 2007; Varma & Madras, 2006). However, saturated fatty acids have a slightly lower reactivity than the unsaturated fatty acids (Warabi et al., 2004).

On the other hand, a second-order kinetic model with respect to both the triglycerides and alcohol concentrations has also been proposed (Diasakou et al., 1998; Song et al., 2008). This divides the transesterification reaction into three steps; the reaction between a triglyceride and an alcohol that generates a diglyceride and an alkyl ester, the diglyceride and alcohol and finally the monoglyceride and alcohol. The concentration of the intermediates is then taken into account and the rate constants are found by mathematical model fitting. Thus, although more complex, the second-order kinetic model is more appropriate than the first-order model for reactions involving alcohol to oil molar ratios below 24:1.

2.2.2 The phase behavior of reactants in supercritical transesterification

The fact that the required optimal operating parameters can become milder with the addition of co-solvents has spurred much interest in the phase behavior of reactants during supercritical transesterification. The phase behavior of soybean oil-methanol with propane as a co-solvent was reported first (Hegel et al., 2007), followed by that for soybean oil-methanol and soybean oil-ethanol with carbon dioxide as the co-solvent (Anitescu et al., 2008). The study of the phase behavior of supercritical transesterification, when performed in a high-pressure view cell, revealed that the liquid - liquid (LL) alcohol - triglycerides mixture transforms to a vapor – liquid - liquid (VLL) phase equilibrium. The VLL equilibrium consists of two immiscible liquid phases (triglycerides and alcohol) and a vapor phase which mainly contains alcohol. Then, the VLL equilibrium changes to a vapor - liquid (VL) phase as a result of the triglycerides dissolving into the supercritical alcohol phase. Finally, the VL equilibrium merges to a one-phase supercritical at nearly the estimated critical point of mixture.

The transition temperature of the VLL to VL equilibriums decreases with increasing methanol to oil molar ratios (Anitescu et al., 2008; Hegel et al., 2007). For example, the reaction mixture of soybean oil and methanol are partially miscible up to temperatures close to 350 °C at a methanol to oil molar ratio of 24:1, while the two liquid phases of soybean oil and methanol become completely miscible at 180 °C and 157 °C with a methanol to oil molar ratio of 40:1 and 65:1, respectively. For a soybean oil-ethanol mixture, it becomes a VL equilibrium at a lower temperature than that for the soybean oil-methanol due to the higher solubility of soybean-oil in ethanol than in methanol (Anitescu et al., 2008).

The transition from a VL system to a one-phase supercritical system was observed near the estimated critical temperature of the mixture, as described in Section 1.1. At a methanol to oil molar ratio of 24:1, the critical temperature of the soybean oil-methanol mixture was 377 °C where the transition temperature was reported to be higher than 350 °C (Anitescu et al., 2008). Moreover, the transition temperature of the two-phase VL to a one-phase supercritical could be reduced by the addition of gaseous co-solvents, such as carbon dioxide and propane. For instance, the addition of 24% by weight of propane decreased the transition temperature of soybean oil-methanol, at a methanol to oil molar ratio of 65:1, from 315 °C to 243 °C (Hegel et al., 2007). Therefore, the addition of gaseous co-solvents is able to reduce the original severe conditions due to their ability to lower the transition temperature from a VL system to a one-phase supercritical system.

2.3 The advantages and drawbacks of supercritical transesterification

Novel solid heterogeneous catalysts that catalyze the transesterification on acidic or basic surfaces instead of in solution have been proposed to overcome the drawbacks of the conventional homogeneous catalytic method, which in part are the same as the supercritical transesterification. The enzyme catalysts typically also allow for a very high selectivity on the alkyl ester products (Helwani et al., 2009; Lene et al., 2009).

2.3.1 The advantages of supercritical transesterification

The advantages of supercritical transesterification over the conventional homogeneous catalytic method are feedstock flexibility, higher production efficiency and it is more environmentally friendly. The feedstock quality is far less influential under supercritical conditions than with the heterogeneous catalytic method, whilst supercritical transesterification has a similar advantage with respect to the product separation as the novel catalytic methods, but it has a higher production efficiency than both novel catalytic methods. The feedstock flexibility is the most important advantage to consider for biodiesel production methods because the resultant biodiesel price strongly depends on the feedstock price (Kulkarni & Dalai, 2006; Lam et al., 2010). The free fatty acids and moisture in low-grade feedstocks and hydrated ethanol pose a negative effect on the basic homogeneous and heterogeneous catalytic methods. Whereas, free fatty acid levels and moisture contents in the feedstock do not significantly affect supercritical transesterification with methanol or ethanol. Therefore, supercritical transesterification is more suitable for use with the low-grade and/or the hydrated ethanol feedstocks (Demirbas, 2009; Gui et al., 2009; Kusdiana & Saka, 2004b; Vieitez et al., 2011). For example, the in-situ transesterification of wet algal biomass in supercritical ethanol gave a 100% alkyl ester yield (Levine et al., 2010).

Supercritical transesterification has a better production efficiency than the conventional catalytic method because it requires a smaller number of processing steps. For instance, the feedstock pretreatment to remove moisture and free fatty acids, and the post-production product treatment steps, such as neutralization, washing and drying, are not necessary. In addition, the rate of reaction under supercritical conditions is significantly faster than the conventional catalytic method, so that the supercritical transesterification requires a smaller reactor size for a given production output.

With respect to environmental aspects, supercritical transesterification does not require any catalysts or chemicals, whilst the waste from the pretreatment and post-treatment steps are also reduced, since those steps are not necessary, leading to the generation of insignificant waste levels. However, the distillation process to recover the excess alcohol requires a large amount of energy which reduces the environmentally friendly advantage of the process (Kiwjaroun et al., 2009). Thus to maintain an environmentally friendly advantage, low-energy separation methods, such as medium pressure flash drum, must be applied to recover the excess alcohol (Diaz et al., 2009).

2.3.2 The drawbacks of supercritical transesterification

The original parameters to achieve a high transesterification conversion were a high temperature (330 – 350 °C), high pressure (19 – 35 MPa) and high alcohol to oil molar ratio (1:40 – 1:42). Indeed, the high temperature and pressure requires both an expensive reactor and a sophisticate energy and safety management policy. As a result of the high alcohol to oil molar ratio a large energy consumption in the reactants pre-heating and recycling steps is required. Moreover, the high amount of alcohol in the biodiesel product retards the biodiesel-glycerol phase separation. Therefore, the use of those original parameters results in high capital costs, especially for the reactor and pump, being somewhat higher than the novel catalytic methods.

To increase the technical and economical feasibility of supercritical transesterification, further studies are required to reduce the energy consumption and operating parameters of this process. For example, the integration of a heating and cooling system can improve (reduce) the energy demand. The experimental techniques that have demonstrated the ability to lower the original parameters for supercritical transesterification are illustrated in sections 3.2 – 3.4.

2.4 The economical feasibilities and chemical limitations of supercritical transesterification

The economical feasibilities of supercritical transesterification, compared with the conventional homogeneous catalytic methods, have been studied by computer simulation (van Kasteren & Nisworo, 2007). These studies usually employed the original parameters for transesterification in supercritical methanol (350 °C, 20.0 MPa and 1:42 methanol to oil molar ratio) and the general parameters for the conventional catalytic methods, such as a reaction at 60 °C, 0.1 MPa and a methanol to oil molar ratio of 9:1. With respect to chemical limitations, supercritical transesterification is limited by the operating temperature due to the thermal cracking of the unsaturated fatty acids.

2.4.1 The economical feasibilities of supercritical transesterification

Supercritical transesterification with the original reacting parameters is economically competitive compared to the conventional catalytic method especially when low-grade feedstocks are employed (van Kasteren & Nisworo, 2007). As expected, the supercritical transesterification has a larger capital cost, due to the required reacting and pumping systems, than the conventional catalytic method, but has no additional capital and operating costs on feedstock pre-treatment, product post-treatment and waste management.

For a better economic feasibility, research into ways to reduce the high operating conditions and lower energy consumption are warranted. For example, supercritical transesterification is not economically feasible when the heating and cooling integration is not employed (Marchetti & Errazu, 2008), while it is a feasible method when heat integration and the presence of catalysts are applied (D'Ippolito et al., 2006; Glišic et al., 2009; van Kasteren & Nisworo, 2007). The addition of calcium oxide as a solid catalyst and the reduction of the alcohol to oil molar ratio significantly decreased the total energy demand and improved the economic feasibility as well (Glišic & Skala, 2009). Furthermore, the addition of propane as co-solvent also enhanced the economic feasibility of supercritical transesterification with methanol (van Kasteren & Nisworo, 2007). However, additional feasibility studies on other assisting techniques that lower the original parameters (see Sections 3.2 - 3.4) are still required.

2.4.2 The chemical limitation of supercritical transesterification

To fulfill the international standard of biodiesel (EN14214), which requires over 96.5% esters content, thermal cracking of polyunsaturated fatty acids is a serious obstacle. At over 300 °C and a reaction time of over 15 min, the methyl linoleate content in biodiesel decreases by approximately 10% compared with the level in the feedstock (Quesada-Medina & Olivares-Carrillo, 2011). Whereas, the % recovery of biodiesel samples which are prepared from various vegetable oils remains constant after exposure with supercritical methanol at 270 °C over 40 min (Imahara et al., 2008). Therefore, the 96.5% alkyl esters content requirement for biodiesel cannot be achieved when an operating temperature of over 300 °C and a reaction time of over 15 min are employed.

To prevent this thermal degradation, the suggested temperature for supercritical transesterification is below 300 °C, and preferably 270 °C. However, the required reaction time to nearly complete transesterification conversion at a 42:1 alcohol to oil molar ratio is then significantly longer at more than 90 min (Minami & Saka, 2006). This prolonged reaction time might cause a decline in the production efficiency obtained by supercritical transesterification, but it could be shortened by the use of assisting methods, as discussed in Sections 3.2 – 3.4.

On the other hand, the gradual heating technique in a tubular reactor has been demonstrated to avoid the thermal cracking of unsaturated fatty acids and shorten the reaction time at the same time (He et al., 2007b). For instance, when the reaction mixture is heated in a tubular reactor gradually from 100 °C at the inlet to 320 °C at the outlet, the biodiesel product obtained after 25 min of reaction time has an over 96% methyl ester content (He et al., 2007b).

3. Process improvements

The process improvements to the supercritical transesterification can be divided into three routes; the addition of the co-solvents, the use of catalysts and process modifications. The general goal, to reduce the original parameters altogether, is the most challenging aspect for supercritical transesterification. The reduced parameters are 270 – 300 °C, 15 – 20 MPa and an alcohol to oil molar ratio of 24:1 – 35:1.

3.1 The chronological development of supercritical transesterification

In 1998, non-catalytic transesterification of soybean oil at the near-critical point of methanol (230 °C, 6.2 MPa and a 27:1 methanol to oil molar ratio) was invented as an alternative method to produce biodiesel, but this method obtained only an 85% methyl ester content after over 10 hours (Diasakou et al., 1998). In 2001, the pioneering transesterification of rapeseed oil in supercritical methanol at 350 °C, 45 MPa and a 42:1 methanol to oil molar ratio, attaining a high methyl ester content (98%) after only 4 min was reported (Kusdiana & Saka, 2001; Saka & Kusdiana, 2001). Transesterification in supercritical methanol has evolved continuously since 2001.

In 2002, the transesterification of cottonseed, hazelnut kernel, poppy seed, safflower and sunflower derived oils in supercritical methanol were evaluated, with a nearly complete transesterification reaction being found for all of the vegetable oils (Demirbas, 2002). Meanwhile, the effect of water and free fatty acids (Kusdiana & Saka, 2004b), the catalytic effect of a metal reactor for supercritical transesterification with methanol (Dasari et al.,

2003; Kusdiana & Saka, 2004a) and the reactivity of supercritical alcohols were all reported (Warabi et al., 2004). In 2004, the first supercritical transesterification of sunflower oil with ethanol and supercritical carbon dioxide in the presence of a lipase enzyme were investigated in a batch reactor (Madras et al., 2004). However, during 2001 – 2005, the maximum alkyl ester contents were generally observed at nearly the same reaction conditions as that reported earlier by the Japanese pioneers (Kusdiana & Saka, 2001; Saka & Kusdiana, 2001).

In 2005, carbon dioxide and propane were introduced as co-solvents to obtain milder operating parameters for the supercritical transesterification with methanol (Cao et al., 2005; Han et al., 2005). Then, the two-step supercritical process (Minami & Saka, 2006) was demonstrated to reduce those operating parameters. In the following years, various catalysts were employed to assist the supercritical transesterification to achieve the maximum alkyl esters content but at milder operating conditions (Demirbas, 2007; Wang et al., 2008; Wang et al., 2007; Wang & Yang, 2007; Yin et al., 2008b). The continuous production of biodiesel in supercritical methanol was reported in 2006 (Bunyakiat et al., 2006) (Minami & Saka, 2006) and 2007 (He et al., 2007b). Therefore, the research focus on the reduction of the elevated operating conditions and continuous process has been ongoing since 2005.

In 2007, the gradual heating technique was introduced to limit or prevent thermal cracking of the unsaturated fatty acids and so prevent the reduction in the final methyl esters content obtained (He et al., 2007b). At the same time, the effect of using co-solvents to reduce the viscosity of vegetable oils was successfully investigated (Sawangkeaw et al., 2007). Supercritical transesterification in ethanol was studied in a continuous reactor in 2008 (Vieitez et al., 2008). In 2009, carbon dioxide was applied to supercritical transesterification with ethanol to reduce the operating conditions (Bertoldi et al., 2009). From 2007 to 2010, numerous additional studies, such as vapor-liquid equilibria of binary systems (Anitescu et al., 2008; Fang et al., 2008; Shimoyama et al., 2008; Shimoyama et al., 2009; Tang et al., 2006), phase behavior of the reaction mixture (Glišic & Skala, 2010; Hegel et al., 2008; Hegel et al., 2007), thermal stability of unsaturated fatty acids in supercritical methanol (Imahara et al., 2008) and process simulation and economic analysis (Busto et al., 2006; D'Ippolito et al., 2006; Deshpande et al., 2010; Diaz et al., 2009; van Kasteren & Nisworo, 2007) were reported, leading to a better understanding of the supercritical transesterification process.

3.2 The addition of co-solvents
The co-solvents that have been used in supercritical transesterification are liquid co-solvents, such as hexane and tetrahydrofuran (THF), and gaseous co-solvents, such as propane, carbon dioxide (CO_2) and nitrogen (N_2). Both types of co-solvents have different purposes and advantages that will be presented accordingly.

The liquid co-solvents are added into the supercritical transesterification reaction to reduce the viscosity of the vegetable oils, which might otherwise pose some pumping problems in a continuous process (Sawangkeaw et al., 2007). Since hexane is the conventional solvent for vegetable oil extraction, it could be possible to combine the supercritical transesterification after the extraction process using hexane for both. Additionally, THF improves the solubility of alcohols in the triglyceride and so forms a single phase mixture, allowing a single high-pressure pump to be employed to feed the reaction mixture into the reactor. A small amount of liquid co-solvent, up to ~20% (v/v) of hexane in vegetable oil, neither affects the transesterification conversion nor lowers the original operating parameters. Whereas, an excess amount of hexane shows a negative effect on the final obtained alkyl esters content due to dilution and obstruction of the reactants (Tan et al., 2010a).

The addition of gaseous co-solvents to the supercritical transesterification reaction aims to reduce the original operating parameters. Due to the fact that the critical properties of gaseous co-solvents are much lower than alcohol and triglycerides, the addition of a small amount of gaseous co-solvents dramatically decreases the critical point of the reaction mixture allowing the use of milder operating parameters. For example, 0.10 mole of CO_2 or 0.05 mole of propane per mole of methanol lowers the reaction temperature and methanol to oil molar ratio to 280 °C and 1:24, respectively (Cao et al., 2005; Han et al., 2005). Furthermore, it was reported that the addition of N_2 improved the oxidation stability and reduced the total glycerol content in the biodiesel product (Imahara et al., 2009). Gaseous co-solvents have the advantage of easier separation from the product than the liquid co-solvents. For instance, they can be separated from the biodiesel product by expansion without using additional energy at the end of the transesterification process, unlike the liquid co-solvents that typically need to be recovered by distillation.

3.3 The use of catalysts

The homogeneous acidic and basic catalysts, such as H_3PO_4, NaOH and KOH, have been applied to supercritical transesterification to obtain milder operating conditions (Wang et al., 2008; Wang et al., 2007; Yin et al., 2008b). However, despite the milder operating conditions and faster rate of reaction obtained compared to the catalyst-free process, the addition of homogeneous catalysts is not an interesting idea because the problem of subsequent catalyst separation and waste management still remain, the same situation as with the conventional homogeneous catalytic process. The use of solid heterogeneous catalysts might enhance the technical and economical feasibility of using supercritical transesterification as a result of the ease of separation of the catalysts. However, the acidic and basic heterogeneous catalysts have different characteristics and advantages, as will be discussed below.

The acidic heterogeneous catalysts, such as WO_3/ZrO_2, zirconia-alumina, sulfated tin oxide and Mg–Al–CO_3 hydrotalcites, have been evaluated in the supercritical transesterification process (Helwani et al., 2009). However, despite the presence of the catalysts, the chemical kinetics of the acidic heterogeneous catalysts at atmospheric pressure were slower than the catalyst-free process. For example, the transesterification of soybean oil in supercritical methanol at 250 °C and a 40:1 methanol to oil molar ratio in the presence of WO_3/ZrO_2 as catalyst still takes 20 hours to attain a 90% conversion level (Furuta et al., 2004). However, the acidic catalysts are less sensitive to moisture and free fatty acid content than the basic catalysts and so they could be appropriate for low-grade feedstocks.

Alternatively, basic heterogeneous catalysts, such as CaO (Demirbas, 2007) MgO (Demirbas, 2008) and nano-MgO (Wang & Yang, 2007), have been applied to supercritical transesterification to reduce the original operating conditions. These catalysts have the ability to catalyze the transesterification reaction at the boiling point of alcohols and are stable at supercritical conditions. As expected, the rate of reaction at the supercritical conditions is faster than that at lower temperatures. For example, the CaO catalyst takes over 180 min to reach over 95% conversion at 65 °C (Liu et al., 2008), but only 10 min to reach complete conversion at 250 °C (Demirbas, 2007). Unfortunately, the basic catalysts can be poisoned by the presence of water and free fatty acids. Therefore, further studies on using low-grade feedstocks with basic heterogeneous catalysts are still required.

3.4 The process modifications

The two-step process is based on firstly a hydrolysis reaction in subcritical water to obtain fatty acid products and then secondly the transesterification and esterification reactions in supercritical alcohol to form the alkyl esters product. The two-step process reduces the optimal operating parameters successfully since the hydrolysis and esterification reactions reach complete conversion at a lower temperature than the transesterification reaction does (Minami & Saka, 2006). Nonetheless, the two-step process is more complicated than the single-step process. For example, the process has high-pressure reactors that connect in series with a high-pressure water-glycerol-fatty acid phase separator. Furthermore, the glycerol-water stream, which is contaminated by trace amounts of fatty acids, requires more separation units. Although a distillation tower is the simplest separation unit for handling the glycerol-water stream, it consumes a large amount of energy to operate.

The high-temperature process involves increasing the operating temperature to 400 to 450 °C (Marulanda et al., 2009; Marulanda et al., 2010), so that the operating pressure, methanol to oil molar ratio and reaction time for complete conversion are reduced to 10.0 MPa, 6:1 and 4 min, respectively. As expected, the unsaturated fatty acids are partially consumed by thermal degradation but the oxidation resistance or storage stability of the product might be enhanced. Under these conditions it was reported that triglyceride and glycerol convert to oxygenate liquid fuel with a conversion of up to 99.5%. The glycerol dehydration both increases the fuel yield by up to 10% and reduces the amount of glycerol by-products (Aimaretti et al., 2009). By using the high-temperature process, the simultaneous conversion of triglyceride, free fatty acids and glycerol to liquid fuel is an alternative option that will increase the feasibility and profitability of supercritical transesterification.

4. Process prospective

In this section, the process prospective is split into two on the basis of the operating temperature since the temperature is the key parameter and chemical limitation for supercritical transesterification. The low-temperature approach aims to produce biodiesel that fulfills the 96.5% alkyl esters content requirement for biodiesel, while the high-temperature approach proposes an alternative method to synthesize the biofuel from a triglyceride-base biomass in supercritical conditions.

4.1 The low-temperature approach

The term "Low-temperature approach" defines supercritical transesterification within a temperature range of 270 – 300 °C so as to avoid the thermal degradation of unsaturated fatty acids and to maximize the alkyl esters content in the product. Without the assistance of any co-solvent, catalyst or other process modification techniques, the low-temperature approach employs a high pressure, a high alcohol to oil molar ratio and a long reaction time to achieve the >96.5% alkyl esters content required for biodiesel composition by the international standard. However, with the assisting techniques, as mentioned in Sections 3.2 – 3.4, the optimal conditions of low-temperature approach generally involve 20 – 30 MPa, an alcohol to oil molar ratio of 24:1 and a reaction time over 30 min. The biodiesel product, which typically exceeds the 96.5% alkyl esters content of the international standard for biodiesel (EN14214), can be used as biodiesel.

For future research involving the low-temperature approach, the use of low-grade feedstocks and/or heterogeneous catalysts are very interesting topics. Alternatively, studies on scale up continuous reactors which are more suitable for an industrial scale are required. These have been successfully evaluated in lab-scale tubular reactors (Bunyakiat et al., 2006; He et al., 2007b; Minami & Saka, 2006), but an evaluation on a scaled-up reactor is presently lacking. An optimal reaction time to achieve over 96.5% alkyl esters content is the most important finding for the low-temperature approach studies because it corresponds with reactor sizing and reflects on the economical feasibility.

4.2 The high-temperature approach

The high-temperature approach uses supercritical transesterification at temperatures over 400 °C, as described in Section 3.4. Even though the mono-alkyl esters content in the product from the high-temperature process is always lower than the biodiesel specification value of 96.5%, it can be proposed as an alternative biofuel that would require further studies on engine testing and fuel properties itself. Improved fuel properties, such as the viscosity and density of the biofuel product, from the high-temperature approach have been proposed (Marulanda et al., 2009). Furthermore, the operating temperature and pressure used in the high-temperature approach are close to those for catalytic hydrocracking in conventional petroleum refining, so it has a high possibility that it can be realized in an industrial scale.

Since the high-temperature approach, as recently initiated, has evaluated the triglycerides found in soybean oil (Anitescu et al., 2008) and chicken fat (Marulanda et al., 2009; Marulanda et al., 2010) only, then additional research into other triglycerides are needed. In addition, studies on the economical feasibility and environmental impact are also required. Indeed, the complete fuel properties need examining along with engine testing for the biofuel product for the high-temperature approach (Basha et al., 2009). On the other hand, the fine studies on the reactions pathways and/or chemical kinetics are also attractive works to better understand the high-temperature approach.

5. Conclusion

Supercritical transesterification is a promising method for a more environmentally friendly biodiesel production as a result of its feedstock flexibility, production efficiency and environmentally friendly benefits. For extended details, the review articles on supercritical transesterification with methanol (de Boer & Bahri, 2011; Sawangkeaw et al., 2010), or ethanol (Balat, 2008; Pinnarat & Savage, 2008) and other supercritical technologies (Lee & Saka, 2010; Tan & Lee, 2011) are also available elsewhere.

Even though the knowledgebase of this process has been growing the past decade, more work is still required for an adequate understanding of the process. In spite of its advantage of feedstock flexibility, there has so far been very little research on the use of low-grade feedstocks in supercritical transformation. Furthermore, prospective studies for both the low-temperature and high-temperature approaches, as mentioned previously, are required to realize supercritical transesterification at an industrial scale.

6. Acknowledgments

The authors would like to acknowledge the financial support from Postdoctoral Fellowship (Ratchadaphiseksomphot Endowment Fund) and the Thai Government Stimulus Package 2 (TKK2555), under the Project for Establishment of Comprehensive Center for Innovative Food, Health Products and Agriculture. We also express thanks to Dr. Robert Douglas John Butcher from the Publication Counseling Unit, Faculty of Science, Chulalongkorn University, for English language editing.

7. References

Aimaretti, N., Manuale, D.L., Mazzieri, V.M., Vera, C.R. & Yori, J.C. (2009). Batch Study of Glycerol Decomposition in One-Stage Supercritical Production of Biodiesel. *Energy and Fuels*, Vol.23, No.2, pp. (1076-1080), ISSN 0887-0624

Anitescu, G., Deshpande, A. & Tavlarides, L.L. (2008). Integrated technology for supercritical biodiesel production and power cogeneration. *Energy and Fuels*, Vol.22, No.2, pp. (1391-1399), ISSN 0887-0624

Balat, M. (2008). Biodiesel fuel production from vegetable oils via supercritical ethanol transesterification. *Energy Sources, Part A: Recovery, Utilization and Environmental Effects*, Vol.30, No.5, pp. (429-440), ISSN 1556-7036

Basha, S.A., Gopal, K.R. & Jebaraj, S. (2009). A review on biodiesel production, combustion, emissions and performance. *Renewable and Sustainable Energy Reviews*, Vol.13, No.6-7, pp. (1628-1634), ISSN 1364-0321

Bertoldi, C., da Silva, C., Bernardon, J.P., Corazza, M.L., Filho, L.C., Oliveira, J.V. & Corazza, F.C. (2009). Continuous Production of Biodiesel from Soybean Oil in Supercritical Ethanol and Carbon Dioxide as Cosolvent. *Energy and Fuels*, Vol.23, No.10, pp. (5165-5172), ISSN 0887-0624

Bunyakiat, K., Makmee, S., Sawangkeaw, R. & Ngamprasertsith, S. (2006). Continuous Production of Biodiesel via Transesterification from Vegetable Oils in Supercritical Methanol. *Energy and Fuels*, Vol.20, No.2, pp. (812-817), ISSN 0887-0624

Busto, M., D'Ippolito, S.A., Yori, J.C., Iturria, M.E., Pieck, C.L., Grau, J.M. & Vera, C.R. (2006). Influence of the Axial Dispersion on the Performance of Tubular Reactors during the Noncatalytic Supercritical Transesterification of Triglycerides. *Energy and Fuels*, Vol.20, No.6, pp. (2642-2647), ISSN 0887-0624

Cao, W., Han, H. & Zhang, J. (2005). Preparation of biodiesel from soybean oil using supercritical methanol and co-solvent. *Fuel*, Vol.84, No.4, pp. (347-351), ISSN 0016-2361

D'Ippolito, S.A., Yori, J.C., Iturria, M.E., Pieck, C.L. & Vera, C.R. (2006). Analysis of a Two-Step, Noncatalytic, Supercritical Biodiesel Production Process with Heat Recovery. *Energy and Fuels*, Vol.21, No.1, pp. (339-346), ISSN 0887-0624

Dasari, M., Goff, M. & Suppes, G. (2003). Noncatalytic alcoholysis kinetics of soybean oil. *Journal of the American Oil Chemists' Society*, Vol.80, No.2, pp. (189-192), ISSN 0003-021X

de Boer, K. & Bahri, P.A. (2011). Supercritical methanol for fatty acid methyl ester production: A review. *Biomass and Bioenergy*, Vol.35, No.3, pp. (983-991), ISSN 0961-9534

Demirbas, A. (2002). Biodiesel from vegetable oils via transesterification in supercritical methanol. *Energy Conversion and Management*, Vol.43, No.17, pp. (2349-2356), ISSN 0196-8904

Demirbas, A. (2007). Biodiesel from sunflower oil in supercritical methanol with calcium oxide. *Energy Conversion and Management*, Vol.48, No.3, pp. (937-941), ISSN 0196-8904

Demirbas, A. (2008). Biodiesel from vegetable oils with MgO catalytic transesterification in supercritical methanol. *Energy Sources, Part A: Recovery, Utilization and Environmental Effects*, Vol.30, No.17, pp. (1645-1651), ISSN 1556-7036

Demirbas, A. (2009). Biodiesel from waste cooking oil via base-catalytic and supercritical methanol transesterification. *Energy Conversion and Management*, Vol.50, No.4, pp. (923-927), ISSN 0196-8904

Deshpande, A., Anitescu, G., Rice, P.A. & Tavlarides, L.L. (2010). Supercritical biodiesel production and power cogeneration: Technical and economic feasibilities. *Bioresource Technology*, Vol.101, No.6, pp. (1834-1843), ISSN 0960-8524

Diasakou, M., Louloudi, A. & Papayannakos, N. (1998). Kinetics of the non-catalytic transesterification of soybean oil. *Fuel*, Vol.77, No.12, pp. (1297-1302), ISSN 0016-2361

Diaz, M.S., Espinosa, S. & Brignole, E.A. (2009). Model-Based Cost Minimization in Noncatalytic Biodiesel Production Plants. *Energy and Fuels*, Vol.23, No.11, pp. (5587-5595), ISSN 0887-0624

Espinosa, S., Fornari, T., Bottini, S.B. & Brignole, E.A. (2002). Phase equilibria in mixtures of fatty oils and derivatives with near critical fluids using the GC-EOS model. *The Journal of Supercritical Fluids*, Vol.23, No.2, pp. (91-102), ISSN 0896-8446

Fang, T., Shimoyama, Y., Abeta, T., Iwai, Y., Sasaki, M. & Goto, M. (2008). Phase equilibria for the mixtures of supercritical methanol + C18 methyl esters and supercritical methanol + α-tocopherol. *The Journal of Supercritical Fluids*, Vol.47, No.2, pp. (140-146), ISSN 0896-8446

Furuta, S., Matsuhashi, H. & Arata, K. (2004). Biodiesel fuel production with solid superacid catalysis in fixed bed reactor under atmospheric pressure. *Catalysis Communications*, Vol.5, No.12, pp. (721-723), ISSN

Glišic, S., Lukic, I. & Skala, D. (2009). Biodiesel synthesis at high pressure and temperature: Analysis of energy consumption on industrial scale. *Bioresource Technology*, Vol.100, No.24, pp. (6347-6354), ISSN 0960-8524

Glišic, S. & Skala, D. (2009). The problems in design and detailed analyses of energy consumption for biodiesel synthesis at supercritical conditions. *The Journal of Supercritical Fluids*, Vol.49, No.2, pp. (293-301), ISSN 0896-8446

Glišic, S.B. & Skala, D.U. (2010). Phase transition at subcritical and supercritical conditions of triglycerides methanolysis. *The Journal of Supercritical Fluids*, Vol.54, No.1, pp. (71-80), ISSN 0896-8446

Gui, M.M., Lee, K.T. & Bhatia, S. (2009). Supercritical ethanol technology for the production of biodiesel: Process optimization studies. *The Journal of Supercritical Fluids*, Vol.49, No.2, pp. (286-292), ISSN 0896-8446

Han, H., Cao, W. & Zhang, J. (2005). Preparation of biodiesel from soybean oil using supercritical methanol and CO_2 as co-solvent. *Process Biochemistry*, Vol.40, No.9, pp. (3148-3151), ISSN 1359-5113

He, H., Sun, S., Wang, T. & Zhu, S. (2007a). Transesterification Kinetics of Soybean Oil for Production of Biodiesel in Supercritical Methanol. *Journal of the American Oil Chemists' Society*, Vol.84, No.4, pp. (399-404), ISSN 0003-021X

He, H., Wang, T. & Zhu, S. (2007b). Continuous production of biodiesel fuel from vegetable oil using supercritical methanol process. *Fuel*, Vol.86, No.3, pp. (442-447), ISSN 0016-2361

Hegel, P., Andreatta, A., Pereda, S., Bottini, S. & Brignole, E.A. (2008). High pressure phase equilibria of supercritical alcohols with triglycerides, fatty esters and cosolvents. *Fluid Phase Equilibria*, Vol.266, No.1-2, pp. (31-37), ISSN 0378-3812

Hegel, P., Mabe, G., Pereda, S. & Brignole, E.A. (2007). Phase Transitions in a Biodiesel Reactor Using Supercritical Methanol. *Industrial and Engineering Chemistry Research*, Vol.46, No.19, pp. (6360-6365), ISSN 0888-5885

Helwani, Z., Othman, M.R., Aziz, N., Kim, J. & Fernando, W.J.N. (2009). Solid heterogeneous catalysts for transesterification of triglycerides with methanol: A review. *Applied Catalysis A: General*, Vol.363, No.1-2, pp. (1-10), ISSN 0926-860X

Hoffmann, M.M. & Conradi, M.S. (1998). Are There Hydrogen Bonds in Supercritical Methanol and Ethanol? *The Journal of Physical Chemistry B*, Vol.102, No.1, pp. (263-271), ISSN 1520-6106

Ilham, Z. & Saka, S. (2009). Dimethyl carbonate as potential reactant in non-catalytic biodiesel production by supercritical method. *Bioresource Technology*, Vol.100, No.5, pp. (1793-1796), ISSN 0960-8524

Imahara, H., Minami, E., Hari, S. & Saka, S. (2008). Thermal stability of biodiesel in supercritical methanol. *Fuel*, Vol.87, No.1, pp. (1-6), ISSN 0016-2361

Imahara, H., Xin, J. & Saka, S. (2009). Effect of CO_2/N_2 addition to supercritical methanol on reactivities and fuel qualities in biodiesel production. *Fuel*, Vol.88, No.7, pp. (1329-1332), ISSN 0016-2361

Khuwijitjaru, P., Fujii, T., Adachi, S., Kimura, Y. & Matsuno, R. (2004). Kinetics on the hydrolysis of fatty acid esters in subcritical water. *Chemical Engineering Journal*, Vol.99, No.1, pp. (1-4), ISSN 1385-8947

Kiwjaroun, C., Tubtimdee, C. & Piumsomboon, P. (2009). LCA studies comparing biodiesel synthesized by conventional and supercritical methanol methods. *Journal of Cleaner Production*, Vol.17, No.2, pp. (143-153), ISSN 0959-6526

Krammer, P. & Vogel, H. (2000). Hydrolysis of esters in subcritical and supercritical water. *The Journal of Supercritical Fluids*, Vol.16, No.3, pp. (189-206), ISSN 0896-8446

Kulkarni, M.G. & Dalai, A.K. (2006). Waste Cooking Oil An Economical Source for Biodiesel:‰ A Review. *Industrial and Engineering Chemistry Research*, Vol.45, No.9, pp. (2901-2913), ISSN 0888-5885

Kusdiana, D. & Saka, S. (2001). Kinetics of transesterification in rapeseed oil to biodiesel fuel as treated in supercritical methanol. *Fuel*, Vol.80, No.5, pp. (693-698), ISSN 0016-2361

Kusdiana, D. & Saka, S. (2004a). Catalytic effect of metal reactor in transesterification of vegetable oil. *Journal of the American Oil Chemists' Society*, Vol.81, No.1, pp. (103-104), ISSN 0003-021X

Kusdiana, D. & Saka, S. (2004b). Effects of water on biodiesel fuel production by supercritical methanol treatment. *Bioresource Technology*, Vol.91, No.3, pp. (289-295), ISSN 0960-8524

Lam, M.K., Lee, K.T. & Mohamed, A.R. (2010). Homogeneous, heterogeneous and enzymatic catalysis for transesterification of high free fatty acid oil (waste cooking oil) to biodiesel: A review. *Biotechnology Advances*, Vol.28, No.4, pp. (500-518), ISSN 0734-9750

Lee, J.-S. & Saka, S. (2010). Biodiesel production by heterogeneous catalysts and supercritical technologies. *Bioresource Technology*, Vol.101, No.19, pp. (7191-7200), ISSN 0960-8524

Lene, F., Knud, V.C. & Birgir, N. (2009). A review of the current state of biodiesel production using enzymatic transesterification. *Biotechnology and Bioengineering*, Vol.102, No.5, pp. (1298-1315), ISSN 1097-0290

Levine, R.B., Pinnarat, T. & Savage, P.E. (2010). Biodiesel Production from Wet Algal Biomass through in Situ Lipid Hydrolysis and Supercritical Transesterification. *Energy and Fuels*, Vol.24, No.9, pp. (5235-5243), ISSN 0887-0624

Lima, D.G., Soares, V.C.D., Ribeiro, E.B., Carvalho, D.A., Cardoso, É.C.V., Rassi, F.C., Mundim, K.C., Rubim, J.C. & Suarez, P.A.Z. (2004). Diesel-like fuel obtained by pyrolysis of vegetable oils. *Journal of Analytical and Applied Pyrolysis*, Vol.71, No.2, pp. (987-996), ISSN 0165-2370

Liu, X., He, H., Wang, Y., Zhu, S. & Piao, X. (2008). Transesterification of soybean oil to biodiesel using CaO as a solid base catalyst. *Fuel*, Vol.87, No.2, pp. (216-221), ISSN 0016-2361

Madras, G., Kolluru, C. & Kumar, R. (2004). Synthesis of biodiesel in supercritical fluids. *Fuel*, Vol.83, No.14-15, pp. (2029-2033), ISSN 0016-2361

Marchetti, J.M. & Errazu, A.F. (2008). Technoeconomic study of supercritical biodiesel production plant. *Energy Conversion and Management*, Vol.49, No.8, pp. (2160-2164), ISSN 0196-8904

Marulanda, V.F., Anitescu, G. & Tavlarides, L.L. (2009). Biodiesel Fuels through a Continuous Flow Process of Chicken Fat Supercritical Transesterification. *Energy and Fuels*, Vol.24, No.1, pp. (253-260), ISSN 0887-0624

Marulanda, V.F., Anitescu, G. & Tavlarides, L.L. (2010). Investigations on supercritical transesterification of chicken fat for biodiesel production from low-cost lipid feedstocks. *The Journal of Supercritical Fluids*, Vol.54, No.1, pp. (53-60), ISSN 0896-8446

Meher, L.C., Vidya Sagar, D. & Naik, S.N. (2006). Technical aspects of biodiesel production by transesterification--a review. *Renewable and Sustainable Energy Reviews*, Vol.10, No.3, pp. (248-268), ISSN 1364-0321

Minami, E. & Saka, S. (2006). Kinetics of hydrolysis and methyl esterification for biodiesel production in two-step supercritical methanol process. *Fuel*, Vol.85, No.17-18, pp. (2479-2483), ISSN 0016-2361

Pinnarat, T. & Savage, P.E. (2008). Assessment of noncatalytic biodiesel synthesis using supercritical reaction conditions. *Industrial and Engineering Chemistry Research*, Vol.47, No.18, pp. (6801-6808), ISSN 0888-5885

Poling, B.E., Prausnitz, J.M. & O'connell, J.P. (2001). *The properties of gases and liquids*. (5 ed). McGraw-Hill, ISBN 7011-6822, New York.

Quesada-Medina, J. & Olivares-Carrillo, P. (2011). Evidence of thermal decomposition of fatty acid methyl esters during the synthesis of biodiesel with supercritical methanol. *The Journal of Supercritical Fluids*, Vol.56, No.1, pp. (56-63), ISSN 0896-8446

Rathore, V. & Madras, G. (2007). Synthesis of biodiesel from edible and non-edible oils in supercritical alcohols and enzymatic synthesis in supercritical carbon dioxide. *Fuel*, Vol.86, No.17-18, pp. (2650-2659), ISSN 0016-2361

Saka, S. & Isayama, Y. (2009). A new process for catalyst-free production of biodiesel using supercritical methyl acetate. *Fuel*, Vol.88, No.7, pp. (1307-1313), ISSN 0016-2361

Saka, S. & Kusdiana, D. (2001). Biodiesel fuel from rapeseed oil as prepared in supercritical methanol. *Fuel*, Vol.80, No.2, pp. (225-231), ISSN 0016-2361

Sawangkeaw, R., Bunyakiat, K. & Ngamprasertsith, S. (2007). Effect of co-solvents on production of biodiesel via transesterification in supercritical methanol. *Green Chemistry*, Vol.9, No.6, pp. (679-685), ISSN 1463-9262

Sawangkeaw, R., Bunyakiat, K. & Ngamprasertsith, S. (2010). A review of laboratory-scale research on lipid conversion to biodiesel with supercritical methanol (2001-2009). *The Journal of Supercritical Fluids*, Vol.55, No.1, pp. (1-13), ISSN 0896-8446

Shimoyama, Y., Abeta, T. & Iwai, Y. (2008). Prediction of vapor-liquid equilibria for supercritical alcohol + fatty acid ester systems by SRK equation of state with Wong-Sandler mixing rule based on COSMO theory. *The Journal of Supercritical Fluids*, Vol.46, No.1, pp. (4-9), ISSN 0896-8446

Shimoyama, Y., Abeta, T., Zhao, L. & Iwai, Y. (2009). Measurement and calculation of vapor-liquid equilibria for methanol + glycerol and ethanol + glycerol systems at 493-573 K. *Fluid Phase Equilibria*, Vol.284, No.1, pp. (64-69), ISSN 0378-3812

Silva, C., Weschenfelder, T.A., Rovani, S., Corazza, F.C., Corazza, M.L., Dariva, C. & Oliveira, J.V. (2007). Continuous Production of Fatty Acid Ethyl Esters from Soybean Oil in Compressed Ethanol. *Industrial and Engineering Chemistry Research*, Vol.46, No.16, pp. (5304-5309), ISSN 0888-5885

Song, E.-S., Lim, J.-w., Lee, H.-S. & Lee, Y.-W. (2008). Transesterification of RBD palm oil using supercritical methanol. *The Journal of Supercritical Fluids*, Vol.44, No.3, pp. (356-363), ISSN 0896-8446

Tan, K.T. & Lee, K.T. (2011). A review on supercritical fluids (SCF) technology in sustainable biodiesel production: Potential and challenges. *Renewable and Sustainable Energy Reviews*, Vol.15, No.5, pp. (2452-2456), ISSN 1364-0321

Tan, K.T., Lee, K.T. & Mohamed, A.R. (2010a). Effects of free fatty acids, water content and co-solvent on biodiesel production by supercritical methanol reaction. *The Journal of Supercritical Fluids*, Vol.53, No.1-3, pp. (88-91), ISSN 0896-8446

Tan, K.T., Lee, K.T. & Mohamed, A.R. (2010b). Optimization of supercritical dimethyl carbonate (SCDMC) technology for the production of biodiesel and value-added glycerol carbonate. *Fuel*, Vol.89, No.12, pp. (3833-3839), ISSN 0016-2361

Tang, Z., Du, Z., Min, E., Gao, L., Jiang, T. & Han, B. (2006). Phase equilibria of methanol-triolein system at elevated temperature and pressure. *Fluid Phase Equilibria*, Vol.239, No.1, pp. (8-11), ISSN 0378-3812

van Kasteren, J.M.N. & Nisworo, A.P. (2007). A process model to estimate the cost of industrial scale biodiesel production from waste cooking oil by supercritical transesterification. *Resources, Conservation and Recycling*, Vol.50, No.4, pp. (442-458), ISSN 0921-3449

Varma, M.N. & Madras, G. (2006). Synthesis of Biodiesel from Castor Oil and Linseed Oil in Supercritical Fluids. *Industrial and Engineering Chemistry Research*, Vol.46, No.1, pp. (1-6), ISSN 0888-5885

Velez, A., Hegel, P., Mabe, G. & Brignole, E.A. (2010). Density and Conversion in Biodiesel Production with Supercritical Methanol. *Industrial and Engineering Chemistry Research*, Vol.49, No.16, pp. (7666-7670), ISSN 0888-5885

Vieitez, I., da Silva, C., Borges, G.R., Corazza, F.C., Oliveira, J.V., Grompone, M.A. & Jachmanián, I. (2008). Continuous Production of Soybean Biodiesel in Supercritical Ethanol-Water Mixtures. *Energy and Fuels*, pp., ISSN 0887-0624

Vieitez, I., Pardo, M.J., da Silva, C., Bertoldi, C., de Castilhos, F., Oliveira, J.V., Grompone, M.A. & Jachmanián, I. (2011). Continuous synthesis of castor oil ethyl esters under supercritical ethanol. *The Journal of Supercritical Fluids*, Vol.56, No.3, pp. (271-276), ISSN 0896-8446

W. King, J., L. Holliday, R. & R. List, G. (1999). Hydrolysis of soybean oil in a subcritical water flow reactor. *Green Chemistry*, Vol.1, No.6, pp. (261-264), ISSN 1463-9262

Wang, C.-W., Zhou, J.-F., Chen, W., Wang, W.-G., Wu, Y.-X., Zhang, J.-F., Chi, R.-A. & Ying, W.-Y. (2008). Effect of weak acids as a catalyst on the transesterification of soybean oil in supercritical methanol. *Energy and Fuels*, Vol.22, No.5, pp. (3479-3483), ISSN 0887-0624

Wang, L., He, H., Xie, Z., Yang, J. & Zhu, S. (2007). Transesterification of the crude oil of rapeseed with NaOH in supercritical and subcritical methanol. *Fuel Processing Technology*, Vol.88, No.5, pp. (477-481), ISSN 0378-3820

Wang, L. & Yang, J. (2007). Transesterification of soybean oil with nano-MgO or not in supercritical and subcritical methanol. *Fuel*, Vol.86, No.3, pp. (328-333), ISSN 0016-2361

Warabi, Y., Kusdiana, D. & Saka, S. (2004). Reactivity of triglycerides and fatty acids of rapeseed oil in supercritical alcohols. *Bioresource Technology*, Vol.91, No.3, pp. (283-287), ISSN 0960-8524

Yin, J.-Z., Xiao, M. & Song, J.-B. (2008a). Biodiesel from soybean oil in supercritical methanol with co-solvent. *Energy Conversion and Management*, Vol.49, No.5, pp. (908-912), ISSN 0196-8904

Yin, J.-Z., Xiao, M., Wang, A.-Q. & Xiu, Z.-L. (2008b). Synthesis of biodiesel from soybean oil by coupling catalysis with subcritical methanol. *Energy Conversion and Management*, Vol.49, No.12, pp. (3512-3516), ISSN 0196-8904

5

Transesterification by Reactive Distillation for Synthesis and Characterization of Biodiesel

G.B. Shinde[1], V.S. Sapkal[2], R.S. Sapkal[3] and N.B. Raut[4]

[1]Department of Chemical Engineering,
Sir Visvesvaraya Institute of Technology, Nashik, M.S.,
[2]Sant Tukadoji Maharaj Nagpur University, Nagpur, M.S.,
[3]University Department of Chemical Technology,
Sant Gadgebaba Amravati University, Amravati, M.S.,
[4]Faculty of Engineering, Sohar University, Sultanate of Oman,
[1,2,3]India
[4]Oman

1. Introduction

Rising world fuel prices, the growing demand for energy, and concerns about global warming are the key factors driving renewed interest in renewable energy sources and in bioenergy. Nowadays, the world energy demand has increased significantly due to the global industrialization and increase of population. As a result, the current limited reservoirs will soon be depleted at the current rate of consumption. So, the research in energy focuses on finding an alternative source of energy to the petroleum derived diesel.

India imported about 2/3rd of its petroleum requirement last year, which involved a cost of approximately Rs. 80,000 crores in foreign exchange. Even 5% replacement of petroleum fuel by bio-fuel can help India and save Rs. 4000 corers per year in foreign exchange. It is utmost important that the options for substitution of petroleum fuels be explored to control this import bill. Biodiesel is a suitable substitute for petroleum-derived diesel. It is biodegradable, almost sulfur less and a renewable fuel, though still not produced by environmentally friendly routes. This alternative fuel consists of methyl or ethyl esters, a result of either transesterification of triglycerides (TG) or esterification of free fatty acids (FFAs). Biodiesel fuel has become more attractive because of its environmental benefits, due to the fact that plants and vegetable oils and animal fats are renewable biomass sources.

Currently, most of the biodiesel comes up from transesterification of edible resources such as animal fats, vegetable oils, and even waste cooking oils, under alkaline catalysis conditions. However, the high consumption of catalysts, the formation of soaps, and the low yields, make biodiesel currently more expensive than petroleum-derived fuel. In addition, the plants from which the vegetable oils are produced capture more CO_2 from the atmosphere than the amount that these oils release during their combustion [1].

The three basic routes to biodiesel production from oils and fats are Base catalyzed transesterification of the oil, Direct acid catalyzed transesterification of the oil and conversion of the oil to its fatty acids and then to biodiesel. Out of these three routes the major production of biodiesel is done with the base catalyzed reaction process.

The stoichiometric equation for transesterification reaction [9] in general can be represented as follows:

Overall reaction:

$$\begin{array}{ccc}
R_1COOCH_2 & & HOCH_2 \quad R_1COOCH_3 \\
| & \xrightleftharpoons{\textit{Catalyst}} & | \\
R_2COOCH \quad + \; 3\,CH_3OH & & HOCH \; + \; R_2COOCH_3 \\
| & & | \\
R_3COOCH_2 & & HOCH_2 \quad R_3COOCH_3
\end{array}$$

Triglyceride   *Glycerol   Methyl esters (Biodiesel)*

2. Biodiesel scenarios worldwide

Sr.No	Region/Country	2005	2006	2007	2008	2009
1	North America	6.1	17.1	33.7	45.9	35.2
2	United States	5.9	16.3	32.0	44.1	32.9
3	Central and south America	0.5	2.2	15.2	38.6	57.9
4	Brazil	0.0	1.2	7.0	20.1	27.7
5	Europe	68.1	113.2	137.5	155.0	172.6
6	France	8.4	11.6	18.7	34.4	41.1
7	Germany	39.0	70.4	78.3	61.7	51.2
8	Italy	7.7	11.6	9.2	13.1	13.1
9	Eurasia	0.3	0.3	0.7	2.5	3.8
10	Lithuania	0.1	0.2	0.5	1.3	1.9
11	Asia and Oceania	2.2	9.1	15.8	28.8	38.5
12	China	0.8	4.0	6.0	8.0	8.0
13	India	0.2	0.4	0.2	0.2	0.4
14	Korea South	0.2	0.6	1.7	3.2	5.0
15	Malaysia	0.0	1.1	2.5	4.5	5.7
16	Thailand	0.4	0.4	1.2	7.7	10.5
	WORLD	77.2	142.0	202.9	270.9	308.2

Source- U.S. Energy Information Administration, International Energy Statistics, Biofuels Production

Table 1. World biodiesel productions by region and selected countries 2005-2009 (Thousand barrels per day)

3. Reactive distillation

Reactive distillation is a chemical unit operation in which chemical reaction and product separation occurs simultaneously in one unit. Reactive distillation column consists of a reactive section in the middle with non-reactive rectifying and stripping sections at the top and bottom.

Let us begin by considering a reversible reaction scheme where A and B react to give C and D. The boiling point of the components follows the sequence A, C, D and B. The traditional flow sheet for this process consists of a reactor followed by a sequence of distillation columns. The mixture of A and B is fed to the reactor, where the reaction takes place in the presence of a catalyst and reaches equilibrium. A distillation train is required to produce pure products C and D. The unreacted components, A and B, are recycled back to the reactor.

The Reactive distillation technology offers many benefits as well as restrictions over the conventional process of reaction followed by distillation or other separation approaches. Reducing capital cost, higher conversion, improving selectivity, lower energy consumption, the reduction or elimination of solvents in the process and voidance of azeotropes   are a few of the potential advantages offered by Reactive distillation. Conversion can be increased far beyond what is expected by the equilibrium due to the continuous removal of reaction products from the reactive zone. This helps to reduce capital and investment costs and may be important for sustainable development due to a lower consumption of resources.[7]

The fig.1 represents the general configuration of reactive distillation.

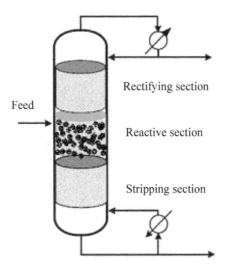

Fig. 1. The general configuration of Reactive Distillation

Based on the applied separation technology, reactive distillation, reactive extraction, reactive adsorption and other combined processes have been distinguished. The combined simultaneous performance of chemical reaction and a multi-component distillation process is an alternative, which has been increasingly used for the large-scale production of relevant chemicals. The use of reactive distillation process can have several advantages such as reduced downstream processing, utilization of heat of reaction for evaporation of liquid phase, simple temperature control of reactor, possibility of influencing chemical equilibria by removal of products and limitations imposed by azeotropic mixture. Several commercially important processes in reactive distillation have been identified in some recent reviews. [7]

Reactive distillation has been successfully applied for the etherification reaction to produce fuel ethers such as methyl tert-butyl ether (MTBE), tert-amyl methyl ether(TAME) and ethyl tertbutyl ether (ETBE). These have been the model reactions for the studies in reactive distillation in the last two decades. A small number of industrial applications of reactive distillation have been around for many decades. Low chemical equilibrium constants can be overcome and high conversions achieved by the removal of products from the location where the reaction is occurring. [6]

It may be advantageous for liquid-phase reaction systems when the reaction must be carried out with a large excess of one or more of the reactants, when a reaction can be driven to completion by removal of one or more of the products as they are formed, or when the product recovery or by-product recycle scheme is complicated or made infeasible by azeotrope formation. Novel processes were proposed based on catalytic reactive distillation and reactive absorption to biodiesel production from esterification and transesterification reactions. The major benefits of this approach were: investment costs reducing about 45% energy savings compared to conventional reactive distillation, very high conversions, increased unit productivity, no excess of alcohol required and no catalyst neutralization step The advantage of reactive distillation can be summarized as follows [3]

a. Simplification: From design view point the combinations of reaction system and separation system can lead to significant capital saving.
b. Improved conversion of reactant approaches 100%. This increase in conversion gives a benefit in reduced recycle costs.
c. Improved selectivity: where, removing one of the products from the reaction mixture or maintaining a low concentration of one of the reagents can lead to reduction of the rates of side reactions and hence improved selectivity for the desired products.
d. Significantly reduced catalyst requirement for the same degree of conversion.
e. Avoidance of azeotropes: RD is particularly advantageous when the reactor product is a mixture of species that can form several azeotropes with each other. RD conditions can allow the azeotropes to be "reacted away" in a single vessel.
f. There is a reduced by-product formation.
g. Heat integration benefits: If the reaction is exothermic, the heat of reaction can be used to provide the heat of vaporization and reduce the reboiler duty.
h. Removal of the product from a system at equilibrium will cause more products to form. Therefore reactive distillation is capable to increase the conversion of equilibrium limited reaction.

Biodiesel production by reactive distillation

As the reaction and separation occurs simultaneously in the same unit in reactive distillation, it is attractive in those systems where certain chemical and phase equilibrium conditions exist. Because there are many types of reactions, there are many types of reactive distillation columns. In this section we describe the ideal classical situation, which will serve to outline the basics of reactive distillation. Consider the system in which the chemical reaction involves two reactants (A and B) producing two products (C and D). The reaction takes place in the liquid phase and is reversible.

$$A+B \leftrightarrow C+D$$

The number of the separation steps depends on the number of products, catalysts, solvents as well as reactants which are not converted. The main objective functions to increase process economics are selectivity as well as reaction yield what influences the reactor design.

Usually, each unit operation is typically performed in individual items of equipment, which, when arranged together in sequence, make up the complete process plant. As reaction and separation stages are carried out in discrete equipment units, equipment and energy costs are added up from these major steps. However, this historical view of plant design is now being challenged by seeking for combination of two or more unit operations into the one plant unit [4].

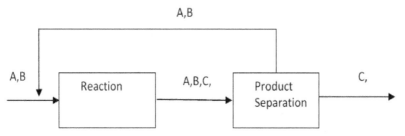

Fig. 2. Standard process scheme for reversible reactions in which the conversion is limited by the chemical equilibrium [9]

For reactive distillation to work, we should be able to remove the products from the reactants by distillation. This implies that the products should be lighter and/or heavier than the reactants. In terms of the relative volatilities of the four components, an ideal case is when one product is the lightest and the other product is the heaviest, with the reactants being the intermediate boiling components.

$$\alpha C > \alpha D > \alpha D$$

The most obvious way to improve the reaction yield in an integrated unit is a continuous separation of one product out of the reaction zone. This allows for getting a 100% conversion in case of reversible reactions [9].

$$A+B \leftrightarrow C+D$$

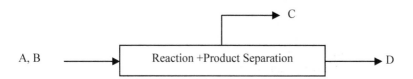

Fig. 3. Complete conversions of reactants in case of equilibrium reaction [7]

Figure 4 presents the flow sheet of this ideal reactive distillation column. In this situation the lighter reactant A is fed into the lower section of the column but not at the very bottom. The heavier reactant B is fed into the upper section of the column but not at the very top. The middle of the column is the reactive section and contains number of reaction trays. The vapor flow rates through the reaction section change from tray to tray because of the heat of the reaction. As component A flows up the column, it reacts with descending B. Very light product C is quickly removed in the vapor phase from the reaction zone and flows up the column. Likewise, very heavy product D is quickly removed in the liquid phase and flows down the column. The section of the column above where the fresh feed of B is introduced (the rectifying section with NR trays) separates light product C from all of the heavier components, so a distillate is produced that is fairly pure product C.

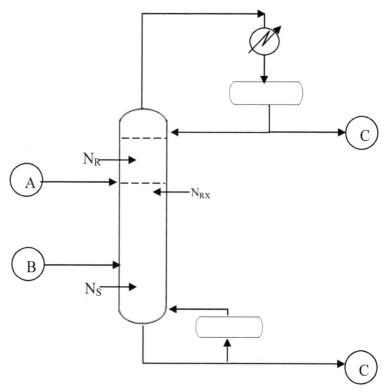

Fig. 4. Flow sheet of ideal reactive distillation column

The section of the column below where the fresh feed of A is introduced (the stripping section with NS trays) separates heavy product D from all of the lighter components, so a bottom is produced that is fairly pure product D. The reflux flow rate and the reboiler heat input can be manipulated to maintain these product purities. The specific numerical case has 30 total trays, consisting of 10 stripping trays, 10 reactive trays, and 10 rectifying trays. Trays are numbered from the bottom. Note that the concentrations of the reactants peak at their respective feed trays. The purities of the two products are both 95 mol%, with B the major impurity in the bottoms and A the major impurity in the distillate [7].

Reactive distillation column must be adjusted to achieve these specifications while optimizing some objective function such as total annual cost (TAC). These design degrees of freedom include pressure, reactive tray holdup, number of reactive trays, location of reactant feed streams, number of stripping trays, number of rectifying trays, reflux ratio, and reboiler heat input [9].

Tray holdup is another design aspect of reactive distillation that is different from conventional. Holdup has no effect on the steady-state design of a conventional column. It certainly affects dynamics but not steady-state design. Column diameter is determined from maximum vapor loading correlations after vapor rates have been determined that achieve the desired separation. Typical design specifications are the concentration of the heavy key component in the distillate and the concentration of the light key component in the bottoms.

However, holdup is very important in reactive distillation because reaction rates directly depend on holdup (or the amount of catalyst) on each tray. This means that the holdup must be known before the column can be designed and before the column diameter is known. As a result, the design procedure for reactive distillation is iterative. A tray holdup is assumed and the column is designed to achieve the desired conversion and product purities. The diameter of the column is calculated from maximum vapor-loading correlations. Then the required height of liquid on the reactive trays to give the assumed tray holdup is calculated. Liquid heights greater than 10–15 cm are undesirable because of hydraulic pressure drop limitations. Thus, if the calculated liquid height is too large, a new and smaller tray holdup is assumed and the design calculations repeated. An alternative, which may be more expensive in terms of capital cost, is to make the column diameter larger than that required by vapor loading [9].

4. Case study - Transesterification by reactive distillation for synthesis and characterization of biodiesel

4.1 Materials and methods

Materials:

a. Oil Feed stocks:
In this study, three commercially available feed stocks of vegetable oils are used .They are
1. Castor seed oil
2. Cottonseed oil
3. Coconut oil

Sample	Kinematic Viscosity, cst (mm²/s)	Density (Kg/m³ at 288K)	Flash point ᵒC	Pour point ᵒC	Saponification value
Castor oil	115 (at 60ᵒC)	938	229	-33	182
Coconut oil	24.85 (at 40 ᵒC)	907	225	20	191.1
Cottonseed oil	35.42 (at 40 ᵒC)	904	15	-15.5	192

Table 2. Physical Properties of Vegetable Oil Feed stocks Used For Transesterification

b. Methanol:
Methanol (Merck) of 99.5% purity (density: 0.785 g/mL at 30ᵒC) was used in this transesterification process.

c. Catalyst:
In this study the catalysts used are:
1. Homogeneous base catalysts (KOH & NaOH)
2. Heterogeneous solid acid catalysts (Amberlyst 15)
The two homogeneous basic catalysts (KOH & NaOH) used for reactive distillation were purchased from local Chemical store at Amravati. M.S.The heterogeneous catalyst used for transesterification Amberlyst BD15 was purchased from Dayo Scientific Laboratory, Nashik Road, Nashik, M.S. India.

Amberlyst-15:

Amberlyst 15 wet is a macro reticular, strongly acidic, polymeric catalyst. Its continuous open pore structure makes it an excellent heterogeneous acid catalyst for a wide variety of organic reactions. Amberlyst 15 is extremely resistant to mechanical and thermal shocks. It also possesses greater resistance to oxidants such as chloride, oxygen and chromates than most other polymeric catalyst. It can use directly in the aqueous system or in organic medium after conditioning with a water miscible solvent. Amberlyst 15 has optimal balance of surface area, acid capacity and pore diameter to make it the catalyst of choice for esterification reactions.

Physical forms	Opaque beads
Ionic form as shipped	Hydrogen
Total exchange capacity	≥1.7 eq /L
Moisture holding capacity	52-57%
Harmonic mean size	600-850 μm
Fine contents	< 0.355 mm :1.0%
Coarse beads	> 1.180 mm :5.0%
Average pore diameter	24 nm
Surface area	45 m^2 / gm
Shrinkage water to methanol	4.0%

Table 3. Characteristics of Amberlyst-15 catalyst

4.2 Transesterification

Transesterification also called alcoholysis is the most common way to produce biodiesel. This involves a catalyzed chemical reaction between vegetable oil and an alcohol to yield fatty acid alkyl esters (i.e., biodiesel) and glycerol. Transesterification is the displacement of alcohol from an ester by another alcohol in a process similar to hydrolysis, except that an alcohol is employed instead of water. Triglycerides, as the main component of vegetable oil, consist of three long chain fatty acids esterified to a glycerol backbone. When triglycerides react with an alcohol (e.g., methanol), the three fatty acid chains are released from the glycerol skeleton and combine with the alcohol to yield fatty acid alkyl esters (e.g., fatty acid methyl esters or biodiesel). Glycerol is produced as a by-product.

The mechanism of transesterification can be represented as follows:

$$
\begin{array}{ll}
CH_2 - OOC - R_1 & R_2 - OOC - R' \quad CH_2 - OH \\
CH - OOC - R_2 + 3R'OH \rightleftarrows & R_2 - OOC - R' + CH_2 - OH \\
CH_2 - OOC - R_1 & R_3 - OOC - R' \quad CH_2 - OH
\end{array}
\tag{1}
$$

$$
\textbf{Triglyceride + Alcohol} \xrightleftharpoons{\text{Catalyst}} \textbf{Esters + Glycerol}
$$

4.2.1 Transesterification of vegetables oils

In the transesterification of different types of oils, triglycerides react with an alcohol, generally methanol or ethanol, to produce esters and glycerin. To make it possible, a catalyst is added to the reaction.The overall process is normally a sequence of three consecutive steps, which are reversible reactions. In the first step, from triglycerides diglyceride is obtained, from diglyceride monoglyceride is produced and in the last step, from monoglycerides glycerin is obtained. In all these reactions esters are produced. The stoichiometric relation between alcohol and the oil is 3:1. However, an excess of alcohol is usually more appropriate to improve the reaction towards the desired product:

$$\text{Triglyceride (TG)} + R'\text{OH} \xrightarrow[k_2]{k_1} \text{Diglycerides (DG)} + R'\text{COOR}_1$$

$$\text{Diglycerides (DG)} + R'\text{OH} \xrightarrow[k_4]{k_3} \text{Monoglycerides (MG)} + R'\text{COOR}_1 \qquad (2)$$

$$\text{Monoglycerides (MG)} + R'\text{OH} \xrightarrow[k_6]{k_5} \text{Glycerin (GL)} + R'\text{COOR}_3 \qquad (3)$$

Startup Procedures of transesterification using reactive distillation:

To start of each experiment, approximate 2 L of oil and 250 mL of methanol were injected into the column. The reboiler heater was set to 120°C and allowed to heat for approximately 1.5 hours till the temperature of the top column reached 62°C.

Steady-operation:

The inputs, both oil at 55°C and methanol at 30°C, were pumped into a short tube mixer to mix the oil with the methanol/catalyst solution. Then the reactant mixture at 62°C was entered to the top of the RD column. In the RD column, triglyceride in the reactant mixture further reacted with the present methanol. The product mixture was withdrawn from the reboiler section and sent to a glycerol ester separator, where the glycerol and esters were separated by gravity in a continuous mode. Every hour, samples were collected from reboiler to analyze the biodiesel composition and methanol content.

In this experimentation reaction parameters has been optimized and an optimized process has been investigated for biodiesel production by transesterification of vegetable oil using reactive distillation technique

Calculations:

The ester content (C) expressed as a fraction in percent, is calculated using the following formula:

$$C = \frac{\left(\sum A\right) - A_{EI}}{A_{EI}} \times \frac{C_{EI} \times V_{EI}}{m} \times 100\% \qquad (4)$$

$\sum A$ = the total peak area from the FAME $C_{14:0}$ to $C_{24:1}$
A_{EI} = the peak area of methyl heptadecanoate
C_{EI} = the concentration , in mg/ml of the methyl heptadecanoate solution
V_{EI} = the volume, in ml of the methyl heptadecanoate solution
m = the mass, in mg of the sample

5. Experimental setup

The system consists of a reactive distillation column fed at the top with the initial reactive solution (oil, alcohol, catalyst). This solution slowly travels down between the plates. When the solution exits the column; the alcohol that has not reacted is recuperated by evaporation. Then, the vapors are re-circulated in the reactive distillation column in the upward direction passing through the plates. As the vapors travel through, interactions between the gaseous alcohol and the liquid solution occur. This then would increase the effective oil to alcohol ratio up to 20:1 (He, Singh et al.2006), thus shifting the reaction equilibrium to the product side and therefore increasing the reaction efficiency. Finally, once the alcohol vapors have reached the top of the reactive distillation column, they are condensed through a condenser allowing the remaining alcohol fraction to re-enter the system. The experimental setup is shown in fig.5 below.

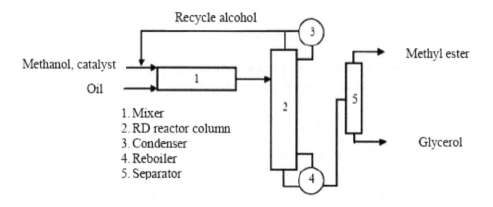

Fig. 5. Schematic of Reactive distillation column for biodiesel

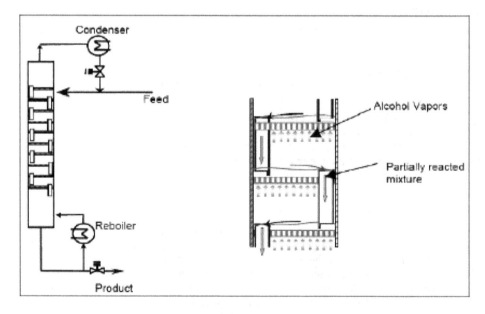

Singh, Thompson Et Al. 2004; Thompson and He 2007

Fig. 6. Operation in Reactive Distillation column

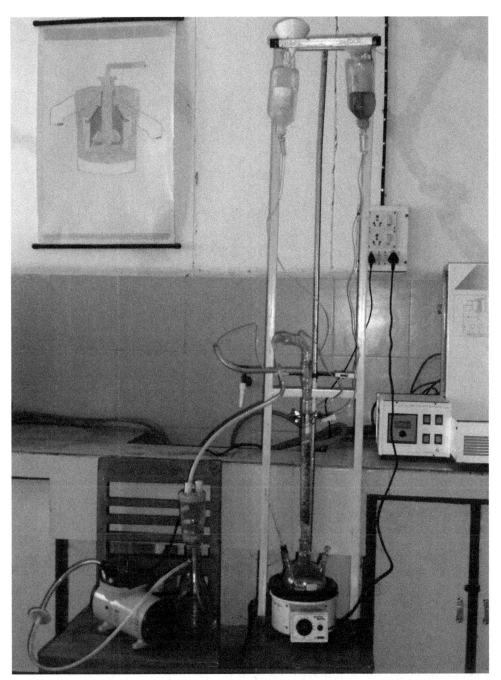

Fig. 7. a) View of experimental lab apparatus of Reactive distillation

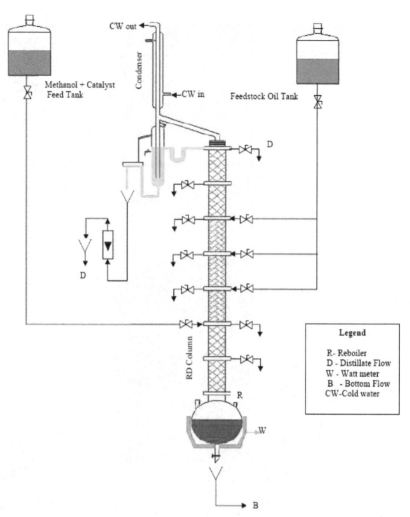

Fig. 7. b) Schematic Diagram of Experimental Setup of Continuous Reactive Distillation Column for biodiesel

In the present experimental study, packed bed Lab-scale reactive distillation column is designed and constructed. This column made up of glass (Inner dia: 30mm, Height of column: 210mm) has been used. The RD column packings used were glass packing. The feed reactants entering into the column were distributed over the packings by the use of distributor plates. The process parameters studied here are alcohol-to-oil ratio.{3:1, 4:1 and 9:1, Optimum methanol-to-oil molar ratio = 4:1},Flow rates of reactants {2, 4, 6 ml/min, Optimum flow rate = 4ml/min},Reaction time {Residence time of 2min, 3min., 6min, optimum residence time = 3min},Temperature {55, 60, 65 °C, Optimum temperature = 65 °C}.

The RD reactor consists of perforated plates or packed sections. For packed columns the packing holds certain amount of reacting liquid in it, forming mini-reactors. Un-reacted alcohol is vaporized from the reboiler, flows upward constantly, and bubbles through the liquid in the packing, which provides a uniform mixing. The thru-vapor is condensed at the top of the RD column and refluxes partially back to the column and the rest combines with the feeding stream

In this study, three non edible vegetable oils namely, castor seed oil, coconut oil and cottonseed oil were used one by one for transesterification.

5.1 Physical and chemical characteristics of the feedstock vegetable oils used for the production of biodiesel

Sample	Kinematic Viscosity, cst(mm²/s)	Density (Kg/m³ at 288K)	Flash point °C	Pour point °C	Saponification value
Castor oil	115 (at 60°C)	938.8	229	-33	182
Cottonseed oil	35.42 (at 40 °C)	904	15	-15.5	192
Coconut oil	24.85 (at 40 °C)	907	225	20	191.1

Table 4. Physical Properties of Vegetable Oil Feed stocks Used for Transesterification:

Oil Sample	Palmitic acid(16:0)	16:1 (Palmitoleic)	Stearic acid(18:0)	Oleic acid (18:1)	Linoleic acid (18:2)	Linolenic acid (18:3)	Other
Castor oil[a]	1.1	0	3.1	4.9	1.3	0	89.6
Cottonseed oil	28.7	0	0.9	13.0	57.4	0	0
Coconut oil	9.7	0.1	3.0	6.9	2.2	0	65.7

Castor oil contains 89.6% ricinoloic acid

Table 5. Table Fatty acids composition of vegetable oils samples under consideration

Physical and chemical characteristics of castor oil

Sr no.	Parameters	Values
1	Appearance	Pale dark yellow
2	Density (at 15°C)	938.8 kg/m³
3	Iodine value	82 -90
4	Saponification value	182
5	Flash Point	229 °C
6	Pour point	-33 °C
7	Acid value	2.0 max
8	Moisture and Volatiles	0.50% max
9	Specific gravity at 20°C	0.954 – 0.967
10	Kinematic Viscosity ,cst(mm²/s)	115 (at 60°C)
	Fatty acids content (%)	
11	Ricinoleic acid	89.6
12	Linoleic acid	4.2%
13	Oleic acid	3%
14	Stearic acid	1%
15	Palmitic acid	1%
16	Linolenic acid	0.3%

Table 6. Physical and chemical characteristics of castor oil

Physical and chemical characteristics of cottonseed oil

Its fatty acid profile generally consists of 70% unsaturated fatty acids including 18% monounsaturated (oleic), 52% polyunsaturated (linoleic) and 26% saturated (primarily palmitic and stearic).

Sr. no.	Parameters	Values
1	Appearance	Golden yellow
2	Density	904
3	Kinematic viscosity(at 40ºC)	35.42 cSt
4	Saponification value	192
5	Flash point	15ºC
6	Pour point	-15.5 ºC
7	Acid value	0.6518 mg of KOH/gm of oil
8	Free fatty acids	0.3258%
9	Molecular weight	863
10	Specific gravity at 20ºC	0.9406 gm/ml

Table 7. Physical and chemical characteristics of cottonseed oil

Sr no.	Parameters	Values
1	Palmitic acid	23%
2	Oleic acid	18%
3	Linoleic acid	52%
4	Aleic acid	19%
5	Alpha lenoleic acid	1%
6	Stearic acid	3%

Table 8. Fatty acids composition of cottonseed oil

Physical and chemical characteristics of coconut oil

Sr no.	Parameters	Values
1	Melting point(ºC)	24
2	Moisture %	<0.1
3	Kinematic viscosity(mm²/s)	24.85 (at 40 ºC)
4	Density (at 15ºC)	907 kg/m³)
3	Iodine value(cgI₂/g)	12-15
5	Saponification value	191.1
6	Flash Point	225 ºC
7	Pour point	20 ºC
8	Total phenolics(mg/kg)	620
Fatty acids content (%)		
9	Saturates	92.0
10	Monosaturates	6.0
11	Polyunsaturates	2.0

Table 9. Physical and chemical characteristics of coconut oil

5.2 Continuous transesterification by reactive distillation for synthesis of biodiesel

The process parameters studied here are alcohol-to-oil ratio.{3:1, 6:1 and 9:1 ,Optimum methanol-to-oil molar ratio =6:1},Reaction time {Residence time of 2min, 3min., 6min, optimum residence time = 3min},Temperature {55, 60, 65°C, Optimum temperature = 65°C }, catalyst loading (1,1.5 and 2 % by wt of oil).

5.2.1 Effect of methanol to oil molar ratio on methyl ester conversion

Feedstock oil was held in a separate heated reservoir maintained at 50°C.The methanol-to-oil molar ratios used were 3.0, 6.0 and 9.0. From several trials, it was found that an overall flow rate of 5-6 ml/min with the column temperature at 64°C provided residence time of about 6min without any significant operational difficulties. The column temperature was maintained by controlling the reboiler heat input. Temperatures above 64°C caused excessive entrainment and a reduction in methanol concentrations in the liquid phase. In preparation for each trial, stock alcoholic KOH was prepared on a stirring plate at a ratio that corresponded to 1, 1.5 and 2 % KOH w/w of oil for each given methanol-to-oil molar ratio, and placed in a holding reservoir next to the RD column. Optimum reaction time in biodiesel formation (1min in prereactor +5min in RD column=6min.). Reaction time by using RD column is 20 times shorter than that in typical batch processes. Also productivity of RD reactor system is 6 to 10 times higher than that of batch and existing continuous flow processes.

The main process parameters examined in this study were as shown below:

For individual oils (Castor, Cottonseed and Coconut oil) under consideration

Methanol/oil molar Ratio (mol/mol)	Methyl esters Conversion (%), Castor oil	Methyl esters Conversion (%), Cottonseed oil	Methyl esters Conversion (%), Coconut oil
3	56	68	55
6	68	72	88
9	72	74	89

Temperature = 64°C, Flow rate =6ml/min, Reaction time = 6min., Catalyst (KOH) =1% by wt. of oil)

Table 10. (a) Effect of Methanol to oil Molar ratio on methyl esters conversion

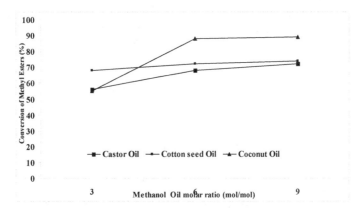

Optimum Molar ratio of Methanol- to- oil= 6:1

Fig. 8. Effect of Methanol to oil Molar ratio on methyl ester conversion

Reaction Time(min.)	Methanol/oil molar Ratio (mol/mol)=3	Methanol/oil molar Ratio (mol/mol)=6	Methanol/oil molar Ratio (mol/mol)=9
2	68	72	55
6	89	96	88
8	91	94	89

Reaction temperature =60°C, Catalyst concentration =1 wt%, Methanol to mixed oil molar ratio = 3:1, 6:1, 9:1

Table 10. (b) Effect of methanol-to-mixed oil feed molar ratio on methyl ester conversion

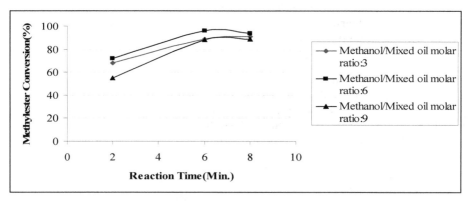

Fig. 9. Effect of Methanol to Mixed oil Molar ratio on methyl ester conversion

The three vegetable oil feedstock's under consideration were mixed and three mixed feed oils were prepared for the experimental runs. The effect of Methanol to Mixed oil Molar ratio on methyl ester conversion was observed as shown in fig.9. The conversion of methyl esters was found to increase with increase in molar ratio during initial reaction time. Also the highest conversion o e 6min in RD column. Thus it can be concluded that the mixed oils can be used for synthesis of biodiesel. This would help in reduction in overall cost of biodiesel synthesis by cutting down the cost of expensive oil by replacing the portion of expensive oils by cheaper oils or the oils which are easily available in abundance.

Reaction Time(min.)	Methyl ester conversion (%) Mixed oil 1	Methyl ester conversion (%) Mixed oil 2	Methyl ester conversion (%) Mixed oil 3
2	65	77	78
6	89	92	95
8	90	93	95

Reaction temperature =60°C, Catalyst concentration =1 wt%, Methanol to mixed oil molar ratio = 6:1, Mixed oil 1= 50% Castor oil+50% Cottonseed oil, Mixed oil 2 = 50% Castor oil+ 50% Coconut oil, Mixed oil 3 = 50% Coconut oil + 50% Cottonseed oil

Table 10. (c) Effect of reaction time using different mixed oils on methyl ester conversion

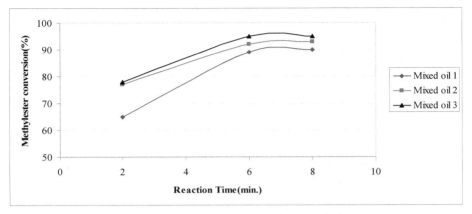

Fig. 10. Effect of reaction time using different mixed oils on methyl ester conversion

It was observed that mixed oil 3 i.e. 50% Coconut oil + 50% Cottonseed oil showed the maximum methyl ester conversion.

5.2.2 Effect of flow rates on methyl ester conversion

Flow rate, ml/min	Methyl esters Conversion (%) , Castor oil	Methyl esters Conversion (%), Cottonseed oil	Methyl esters Conversion (%), Coconut oil
5	90	92	90
6	94	96	93
7	94	95	93

Molar ratio of Methanol to Oil = 6:1, Reaction time = 6min., Catalyst (KOH) =1% (by wt. of oil), Temperature = 60°C

Table 11. Effect of Flow rates on methyl ester conversion

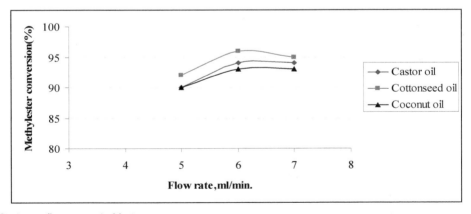

Optimum flow rate = 6ml/min

Fig. 11. Effect of Flow rates on methyl ester conversion

The feed stream flow rates for the test run were chosen carefully in order to avoid any column flooding problem. The flow rate, which is inversely related to retention time, is used as an experimental factor to interpret the reaction conversion with the liquid retention time. The flow rate achieved in the experimental runs varied from 5 to 7 mL/min. The retention time varied from about 4 to 8 min. These values may not be the actual reaction time because of some reaction that takes place in the reboiler. Since the concentrations of methanol and catalyst were very small in the liquid phase of the reboiler, it was not possible to determine the actual retention time of reactants and catalyst in the reboiler. The effect of flow rate was mainly on the production of methyl ester of the reactor. The % weight of methyl ester decreased while the flow rate increased since the retention time is less. For the RD operation of setup in this study, the feed flow rate should not be higher 6mL/min in order to avoid flooding in column and this rate were considered as optimum range of operation.

5.2.3 Effect of reaction time on methyl ester conversion

Reaction Time (min.)	Methyl esters Conversion (%) , Castor oil	Methyl esters Conversion (%), Cottonseed oil	Methyl esters Conversion (%), Coconut oil
4	92	95	89
6	94	96	93
8	94	96	93

Molar ratio of Methanol to Oil = 6:1, Catalyst (KOH) = 1% (by wt. of oil), Flow rate = 6ml/min, Temperature = 60°C

Table 12. Effect of Reaction time on methyl ester conversion

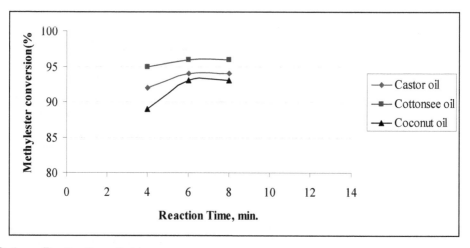

Optimum Reaction time = 6min.

Fig. 12. Effect of reaction time on methyl ester conversion

The rate of transesterification depends on the time of reaction as shown in fig.12 The reaction was slow during the first few minutes due to time taken for mixing and dispersion of methanol with the triglycerides in the oil. However the rate of reaction increased steadily from 6 min of the reaction. The residence time of the reactants in the RD Column was 6min. at which the highest ME conversion was achieved. Thus the reaction time clearly influences the conversion rate of methyl esters.

5.2.4 Effect of catalyst loading on methyl esters conversion

The type of catalyst and the amount of catalyst has a great impact on formation of biodiesel. The adequate catalyst loading is necessary to obtain the maximum conversion of triglycerides to methyl esters. In the above experimentation, different types of catalysts with different catalyst loadings were utilized for transesterification reaction and their effects were studied. In the course of the tests, it is observed that addition of excess amount of catalyst, give rise to the formation of an emulsion, which has increased the viscosity and led to the formation of gel.

a.   Effect of catalyst (KOH) loading on methyl ester conversion

KOH Catalyst loading (wt.%)	Methyl esters Conversion (%), Castor oil	Methyl esters Conversion (%), Cottonseed oil	Methyl esters Conversion (%), Coconut oil
1	80	92	95
1.5	78	93	96
2	60	93	96

Methanol to oil molar ratio = 6, Temperature =60°C, Catalyst loadings used for experimentation = 1% , 1.5% and 2% KOH by wt of oil

Table 13. Effect of catalysts loadings on methyl ester conversion

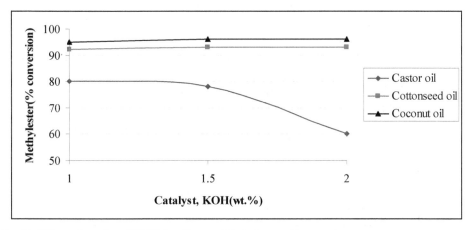

Fig. 13. Effect of catalyst (KOH) loading on Methyl ester Conversion

b.   Effect of catalyst(NaOH) loading on methyl ester conversion

NaOH Catalyst loading (wt.%)	Methyl esters Conversion (%), Castor oil	Methyl esters Conversion (%), Cottonseed oil	Methyl esters Conversion (%), Coconut oil
0.5	72	89	94
1	80	80	95
1.5	78	91	96
2	60	91	97

Methanol to oil molar ratio= 6, Temperature =60°C, Catalyst loadings used for experimentation = 0.5%, 1%, 1.5% and 2% NaOH

Table 14. Effect of Catalyst (NaOH) loadings on Methyl ester Conversion

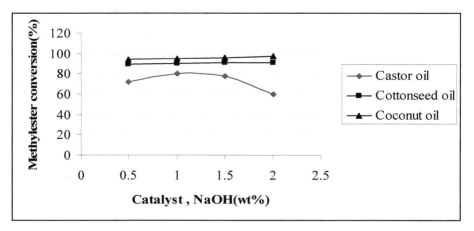

Fig. 14. Effect of catalyst (NaOH) loading on Methyl ester Conversion

It can be seen that the conversion of triglyceride to methyl ester increased as the catalyst loading increased from 0.5 to 1 wt.%. However, with the increase of the catalyst loading from 1% to 2 wt.%, the conversion of triglyceride to methyl ester decreased. Since the conversions of triglycerides to methyl esters did not change significantly for the catalyst loadings from 1wt.%, the optimum catalyst loading for this reaction was 1wt.%.

c.   Effect of catalyst (Amberlyst-15) loading on Methyl ester Conversion

Amberlyst-15 Catalyst loading (wt.%)	Methyl esters Conversion (%), Castor oil	Methyl esters Conversion (%), Cottonseed oil	Methyl esters Conversion (%), Coconut oil
3	89	90	88
4.5	77	80	76
6	56	65	60

Methanol to oil molar ratio= 6, Reaction temperature =60°C, Reaction time = 1.5 hrs

Table 15. Effect of catalyst (Amberlyst-15) loading on Methyl ester Conversion

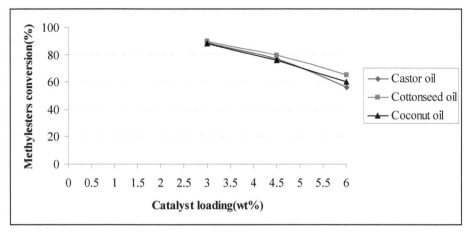

Optimum catalyst loading = 3wt%

Fig. 15. Effect of catalyst (Amberlyst-15) loading on Methyl ester Conversion

It can be observed that there was no rise in ME conversion when catalyst loading was increased from 3 to 6 wt%. The highest conversion was achieved at catalyst loading of 3 wt%.

5.2.5 Effect of reaction time on methyl ester conversion using amberlyst -15 catalyst

Amberlyst-15 a solid acid catalysts been studied for transesterification for production of methyl esters. However, mild reaction conditions are necessary to avoid degradation of the catalyst. At a relatively low temperature (60 °C), the conversion of castor, cottonseed and coconut oil was reported to be only between 2% to 5%, when carrying out the reaction at atmospheric pressure and a 6:1 methanol-to-oil initial molar ratio.

Reaction Time(hrs)	Methyl esters Conversion (%), Castor oil	Methyl esters Conversion (%), Cottonseed oil	Methyl esters Conversion (%), Coconut oil
0.5	60	65	55
1	76	79	74
1.5	80	85	82
3	72	80	76
3.5	65	62	60
4	56	50	52

Methanol to oil molar ratio= 6, Reaction temperature =60°C, Catalyst loading=3wt.% Amberlyst-15

Table 16. Effect of Reaction time on Methyl ester Conversion using Amberlyst -15 Catalyst

During the initial 30 minutes reaction time the percent conversion was less. But as the reaction time progressed the conversion of triglycerides to methyl esters was increased till further one hour. After 1.5 hrs of reaction time, there was no significant rise in conversion. But the ME conversion decreased after 1.5 hrs. So, 1.5 hrs is obtained as the optimum reaction time for this study.

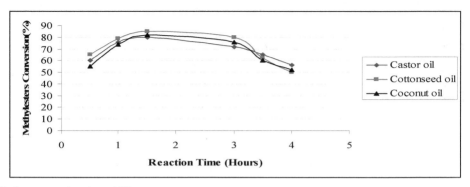

Optimum reaction time =1.5 hours

Fig. 16. Effect of reaction time on Methyl ester Conversion using Amberlyst -15 Catalyst

5.2.6 Effect of methanol to oil molar ratio on methyl ester conversion using Amberlyst-15 catalyst

Methanol-t-oil molar ratio	Methyl esters Conversion (%), Castor oil	Methyl esters Conversion (%), Cottonseed oil	Methyl esters Conversion (%), Coconut oil
6	89	92	90
9	76	84	77
12	44	73	39

Reaction time = 1.5 hrs, Reaction temperature =60°C, Catalyst loading=3wt. % Amberlyst-15

Table 17. Effect of methanol to oil molar ratio on Methyl ester conversion using Amberlyst-15 catalyst

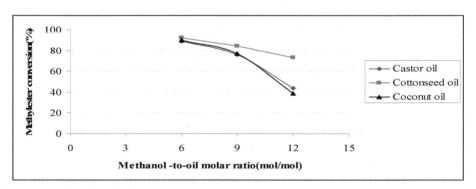

Optimum methanol to oil molar ratio = 6:1

Fig. 17. Effect of methanol to oil molar ratio on methyl ester conversion using Amberlyst-15 catalyst

It can be observed that there was no rise in ME conversion when methanol to oil molar ratio was increased from 6:1 to 12:1 .The highest conversion was achieved at methanol to oil molar ratio of 6:1.So, it is considered as optimum methanol to oil molar ratio.

5.2.7 Evaluation of column operating conditions

Effect of reaction temperature on methyl ester conversion

Temperature (°C)	Methyl esters Conversion (%), Castor oil	Methyl esters Conversion (%), Cottonseed oil	Methyl esters Conversion (%), Coconut oil
55	89	84	90
60	92	96	96
65	94	97	96

Molar ratio of Methanol to Oil = 6:1, Catalyst (KOH) = 1% (by wt. of oil), Flow rate = 6ml/min, reaction time =6min

Table 18. Effect of reaction temperature on methyl ester conversion

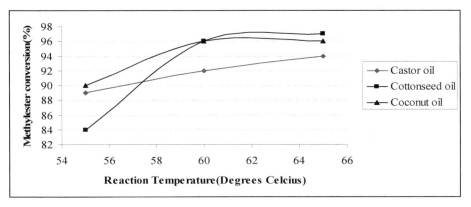

Optimum reaction Temperature = 60°C

Fig. 18. Effect of Temperature on Methyl ester conversion (%)

The effect of temperature on methyl ester conversion was studied by transesterification at different temperature i.e. 55 °C, 60 °C and 65 °C. It was observed that the conversion increased with increase in temperature from 55 °C to 60 °C . But there was no significant rise in conversion after 60 °C upto 65 °C. So, the optimum value of temperature for this transesterification reaction was considered as 60 °C.

5.2.8 Effect of reboiler temperature on methyl ester content of the product

The function of reboiler is to vaporize the residual methanol present in the liquid reaching the bottom of the column. At steady state, the boiling-up rate of methanol is determined by the heat load on the reboiler, heat transfer efficiency and the amount of methanol in the reboiler. Methanol boils at 64.7°C, however, according to the experiments, sufficient methanol vapors were generated only with reboiler temperature higher than 90°C. Depending upon the methanol concentrations, therefore, reboiler temperature in the experimental design varied from 80°C to 120°C in order to produce smooth and consistent methanol vapor flow rates. It was found that the lower reboiler temperatures are favorable for better reactor performance. A possible reason is that with higher operating temperatures, the rates of soap formation increase more rapidly than that of transesterification.

Time (hrs)	%Methyl ester at 120oC	%Methyl ester at 100oC	%Methyl ester at 90oC	%Methyl ester (at 85oC
0	0	0	0	0
1	25	34	10	33
2	44	55	28	63
3	56	67	49	68
4	62	74	68	70
5	66	78	80	72
6	68	80	87	72
8	68	83	90	74
10	66	86	91	75
12	66	88	93	75
14	67	87	92	73
15	67	87	92	74

Molar ratio of Methanol to Oil = 6:1, Flow rate =6ml/min, Reaction time = 6min., Catalyst (KOH) =1% by wt. of oil)

Table 19. Effect of reboiler temperature on methyl ester content of the product

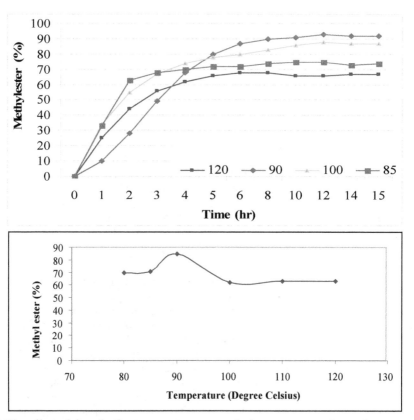

Fig. 19. The effect of the reboiler temperature on methyl esters content in product

5.2.9 Biodiesel yield

The biodiesel yield was determined indirectly from the triglyceride as discussed in the materials and methods section. The percentage composition of the triglyceride consumed was assumed to be converted to methyl ester. Therefore, the amount of biodiesel produced was calculated from the difference in composition of triglyceride in the original oil and final biodiesel.

6. Conclusions

From the experimental investigation done in this research work, the following conclusions can be drawn:

Reactive distillation can be used as a techno economical process for synthesis of biodiesel from vegetable oil by transesterification process. The process proposed here can dramatically improve the economics of current biodiesel synthesis and reduce the number of downstream steps.

Using reactive distillation for synthesis of methyl ester has several advantages such as high unit productivity, up to 6–10 times higher than of the current process, lower excess alcohol requirements, reduced capital and operating costs, due to less units and lower energy consumption, Sulfur-free fuel, since solid acids do not leach into the product, no waste streams because no salts are produced. Also there is significant reduction in reaction time as well as in number of equipment units compared to that in conventional batch and continuous transesterification processes.

The feasibility of using mixture of two different vegetable oils was tested and it was found that coconut oil and cottonseed oil if mixed in equal proportion and used for transesterification reaction, the methyl esters conversion as high as 95% can be achieved.

From the experimental results the optimization of parameters obtained was as follows:

Transesterification reaction is affected by methanol to oil molar ratio, catalyst loading, reaction temperature, flow rate of reactant streams, reaction time, mixing intensity. In the present study reaction was carried out using three values of Methanol to oil molar ratios (3:1, 6:1 and 9:1) for individual oils (castor, cottonseed and coconut oil) and KOH catalyst for synthesis of biodiesel the highest ME conversions obtained for castor, cottonseed and coconut oil transesterification were 68%, 72% and 88% respectively. The highest ME conversion was obtained for coconut oil in this case. Whereas by using the heterogeneous catalyst Amberlyst -15, the highest ME conversions obtained for castor, cottonseed and coconut oil transesterification were 89%, 92% and 90% respectively for methanol to oil molar ratio of 6:1.

Also the possibility of using mixed oils was investigated by using three seed oils in 3 different proportions, such as mixed oil 1 (50% castor oil+50% cottonseed oil), mixed oil 2 (50% castor oil+50% coconut oil) and mixed oil 3(50% coconut oil+50% cottonseed oil).The highest ME conversion (95%) was obtained for mixed oil 3 at methanol to oil molar ratio of 6:1.Thus the option of using mixed oils is feasible in case of scarcity of one of particular feed stock oil in particular region. Also the application can be further investigated to reduce the cost of production of biodiesel by using cheaper feed stocks in more proportion in mixed oil feed.

The flow rate of reactants has an impact on reaction rate of biodiesel production. Out of 5ml/min, 6ml/min and 7ml/min flow rates of reactants to the RD column transesterification reaction. The highest ME conversion (96%) was obtained for cottonseed oil at reactants flow rate of 6ml/min.

For a practical and economic feasible transesterification process, it is necessary to limit the reaction time at a certain period. Longer reaction time could also permit reversible transesterification reaction to occur, which eventually could reduce the yield of fatty acid alkyl esters. Thus, optimization of reaction time is also necessary. In this study, the reaction time was varied from 4 min, 6min and 8min. It was observed that the highest ME conversion (96%) was obtained for cottonseed oil at 6min reaction time. Whereas for heterogeneous catalyst, reaction times of 0.5,1, 1.5, 3, 3.5 and 4 hrs using catalyst Amberlyst-15 for Methanol to oil molar ratio 6:1, Reaction temperature of 60°C, Catalyst loading of 3wt.% were used and the highest ME conversion (85%) was obtained for cottonseed oil after 1.5 hrs reaction time. There was no significant rise in conversion rate after 1.5 hrs.

The catalyst plays an important role in transesterification reaction. The type and quantity of catalyst usually depend upon the quality of feed stock and method applied for transesterification. Three values of catalyst loadings of 1, 1.5 and 2 wt% KOH were used for Methanol to oil molar ratio 6:1 at 60°C .The highest ME conversion (96%) was observed for coconut oil at 1.5 wt% KOH catalyst loading.

Similarly the second homogeneous catalyst NaOH also resulted in the same conversion for the same experimental conditions. Whereas the heterogeneous catalyst Amberlyst-15 was used in three catalyst loadings of 3, 4.5 and 6 wt% Amberlyst-15 for Methanol to oil molar ratio of 6:1, Reaction temperature of 60°C, Reaction time of 1.5 hrs, the highest ME conversion (90%) was obtained for cottonseed oil at 3wt% Amberlyst-15 catalyst loading.

The transesterification was carried out at reaction temperatures of 55, 60 and 65°C for individual oils-castor, cottonseed and coconut oil and the highest ME conversions obtained for castor, cottonseed and coconut oil transesterification were 92%, 96% and 96% respectively for 60°C temperature using KOH catalyst. For homogeneous catalyst, moderate reaction temperature is enough to commence the reaction whereas for heterogeneous catalyst the operating temperature varies depending upon activation energy and conditions to produce the high yield of methyl esters. For NaOH catalyst the same optimum value of 60 °C temperature was obtained in batch transesterification process with maximum ME conversion of 96%.

The sufficient methanol vapors were generated only with reboiler temperature higher than 90°C. Depending upon the methanol concentrations, therefore, reboiler temperature in the experimental design was varied from 80°C to 120°C in order to produce smooth and consistent methanol vapor flow rates. It was found that the lower reboiler temperatures are favorable for better reactor performance. A possible reason is that with higher operating temperatures, the rates of soap formation increase more rapidly than that of transesterification.

It was found that cottonseed oil resulted into maximum yield of biodiesel. Usually crude cottonseed oil contains palmitic acid (22- 26%), oleic acid (15-20%), linoleic acid (49- 58%) and approximately 10% mixture of arachidic acid, behenic acid and lignoceric acid, as well as about 1% sterculic and malvalic acids. In this study, the used crude cottonseed oil contained 24.60% of palmitic acid, 17.09% of oleic acid, and 50.50% of linoleic acid. Since higher amount of free fatty acids (FFA) (>1% w/w) in the feedstock can directly react with the alkaline catalyst to form soaps, which are subject to form stable emulsions and thus prevent separation of the biodiesel from the glycerol fraction and decrease the yield, it is better to select reactant oils with low FFA content or to remove FFA from the oil to an acceptable level before the reaction. Nevertheless, the FFA (calculated as oleic acid) content of the crude cottonseed oil used in this experiment was only 0.8%, which was in an allowed level for being directly used for reaction with the alkaline catalyst to produce biodiesel.

The castor oil showed yield of biodiesel less than that of cottonseed oil due to its high acid value. High acid value leads to neutralization of part of catalyst present thus producing soaps within the reaction medium which reduce mass transfer during the reaction. Also in basic medium, the hydroxyl group at C-12 of ricinoleic acid is converted into an alcohoxide derivative that can compete with the generation of methoxide species and compromise the conversion reaction. The Coconut oil contains approximately 92.1% saturated fatty acids, 6.2% monounsaturated fatty acids, 1.6% polyunsaturated fatty acids. Different fatty acids in coconut oil range from C6 – C18 carbon atom chains. It contains Lauric acid(over 50%) higher in composition So, it was concluded that cottonseed oil was the most feasible feedstock among the three vegetable oil feed stocks under consideration for this study.

The production of biodiesel using transesterification by reactive distillation can be considered as technically as well as economically feasible process and its scale up at industrial level should be recommended to meet the future energy demand.

7. Acknowledgement

First of all, I wish to express my deepest gratitude to my supervisor Hon'ble, **Dr.V.S.Sapkal**, Vice Chancellor, Rashtrasant Tukadoji Maharaj Nagpur University, Nagpur, whose guidance, encouragement, wisdom, motivation, and expectations are indispensable to my achievements and will serve as a continuous inspiration for my future career.

My deepest thanks also go to respected **Dr.R.S.Sapkal**, Head, Department of Chemical Technology, Sant Gadgebaba Amravati University, Amravati, who gave me a lot of helpful ideas, suggestions, and discussions that contribute to the remarkable success achieved in this work.

No thanks are enough to **Dr.N.B.Raut**, Assistant Professor, Faculty of Engineering, Sohar University (Affiliated to Queensland University), Sultanate of Oman, Oman for his valuable direction, motivation and constant support in accomplishment of my research.

I would like to thank **Dr.C.V.Rode, Dr.M.G.Sane**, Senior Scientists from NCL, Pune for their technical support and guidance to my research. Thanks go to all the members in my laboratory as well as librarians of NCL, Pune, SGBAU, Amravati, SVIT, Nashik, KKW College of Engineering, Nashik for their help out in literature survey and analytical facilities in this research.

I also appreciate all my colleagues and other faculties at SVIT, Nashik, **Prof.R.N.Vaidya**, Brahma Valley College of Engineering, Nashik and **Prof.M.G.Shinde**, MET's Institute Of Engineering, BKC, Nashik for their precious direct and indirect assistance in this research work. I would thank lab assistants and staff of UDCT, Amravati for their kind cooperation during the experimentation phases. A special thanks goes to **Mr.P.R.Wankhade** for his valuable support during the experimentation phases at UDCT, Amravati.

Finally, I would like to thank my family and friends for their patience, motivation and admiration. My success is directly related to their love and strong support.

8. References

Chuaohuymak Pojanalai, Sookkumnerd Terasut, Kinetics of Homogeneous Transesterification Reaction of Palm oil and Methanol, Technology p. 1-6,2005.

Demirbas Ayhan, Biodiesel A Realistic Fuel Alternative. Trabzon: Springer-Verlag London Limited, 2008.

Drapcho, Caye, Nghiem Phu Nhuan and Terry H.Walker. Biofuels Engineering Process Technology. New York Chicago San Fr: The McGraw-Hill Companies, Inc, 2008.

Kalayasiri, P., Jayashke, N. and Krisnangkura, K. "Survey of seed oils for use as diesel fuels." Journal of American Oil Chemical Society (1996): 73:471–474.

Noshadi I, A Review of Biodiesel Production Via Reactive Distillation AICHE Conference, Spring Session,2011

Richardson, Colson and. Particle Technology and Separation Processes. Newcastle: Library of Congress Cataloguing in Publication Data, 1991.

Sharma Y.C, Singh Bhaskar, A hybrid feedstock for a very efficient preparation of biodiesel, Technology, Volume, Pages 1267-1273, 2010

Sundmatcher, Kai and Achim Kienle. Reactive Distillation Status and Future Directions. Mhar. Achim Kienle Kai Sundmatcher. Megdeburg: Wiley-VCH, 2002.

Viswanathan B. and A.V.Ramaswamy, 'Selection Of Solid Heterogeneous Catalysts For Transesterification Reaction', National Centre for Catalysis Research, Indian Institute of Technology, Madras, Chennai 600 036

Zhang Y, Dube M A, McLean D D, Kates D, analysis, Bioresource Technology 90 (3) p. 229-240, 2003

http://www.crnindia.com
http://www.castoroil.in
http://www.mcxindia.com
http://www.biodiesel.org
http://www.coconutboard.nic.in
http://www.biodiesel.com
http://www.teriin.org
http://www.indexmundi.com

6

An Alternative Eco-Friendly Avenue for Castor Oil Biodiesel: Use of Solid Supported Acidic Salt Catalyst

Amrit Goswami

Chemical Science Block, CSIR- North-East Institute of Science & Technology,
(A Constituent Establishment of Council of Scientific and Industrial Research,
Govt. of India), Jorhat-785006, Assam,
India

1. Introduction

Year back in 1885, when Rudolf Diesel first invented the diesel engine, it was intended to run it on oil from vegetative sources and in the course of time with gradual depletion of the fossil fuel has now become a mandate of the day. Because it will play an important part in sustainable fuel and energy production solution for the future. Vegetable oil which remains in the form of triglyceride(1) of long chain fatty acid with carbon chain C_{16}-C_{18} is not fit to use directly but needs certain transformation such as pyrolysis, microemulsion formation, transesterification etc. to suit it to use as diesel fuel

$$CH_2OR$$
$$|$$
$$CHOR$$
$$|$$
$$CH_2OR$$

(1)

R= Long Chain Fatty acid moiety

Triglyceride

The transformed oil is termed as biodiesel due to its original biological source. Finite fossil fuel reserve, political, economic, biodegradability, low toxicity, health and environmental issues have led it to consider as the alternative and more importantly renewable and eco-friendly fuel. It has been found to show its ability to meet the world energy demand in transportation along with agricultural and other industrial sectors (Akoh et al. 2007). Since the source is plant, it is green as it does not have ash content, sulphur, aromatic ring compounds, renewable and so it has come out as superlative alternative and can be used in compression ignition engine with minor or no modification of the engine (Xu and Wu, 2005). A breakthrough in the process of converting vegetable oil into useful form promises a cheaper way to go green as it contributes mitigating global warming also. However the slow pace of progress in this direction in alternative fuel technologies has prevented the vision from materializing. On the other hand vegetable oil as such is expensive and direct use of it in diesel engine is not possible. Because, firstly the vegetable oils are very viscous. High viscosity in fuel causes transportation problems, carbon deposits in engine, suffering of engine liner, injection nozzle failure, and gum formation, lubricating oil thickening and high cloud and pour point. Secondly, the glyceride moiety in the triglyceride form of the vegetable oil during combustion could lead to formation of acrolein (2) and this in turn lead to formation of different aromatics (3) as polluting by-products. This is one of the reasons why fatty esters of vegetable oils are preferred over triglycerides.

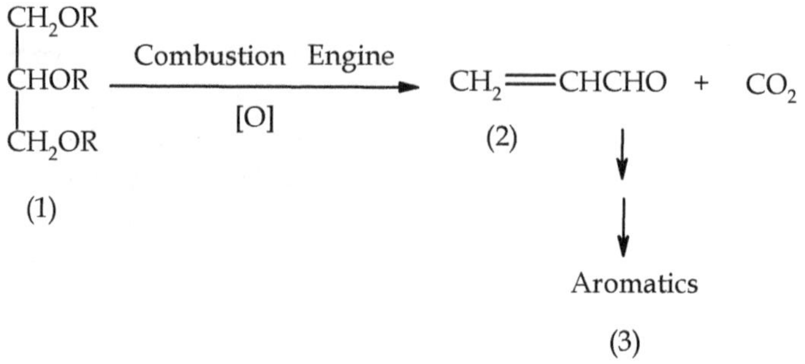

$$
\begin{array}{c}
CH_2OR \\
| \\
CHOR \\
| \\
CH_2OR \\
(1)
\end{array}
\xrightarrow[\text{[O]}]{\text{Combustion Engine}}
CH_2{=}CHCHO \; + \; CO_2
$$

(2) → → → Aromatics

(3)

Scheme 1.

In 1970 it was discovered that reduction of viscosity of vegetable oils could be made by simple chemical process called transesterification by which the vegetable oil is treated with a low alkyl alcohol such as methanol or ethanol in presence of a suitable catalyst to form low alkyl esters whereby it could perform as petro diesel in modern engine. Glycerol that is produced during transesterification as by-product can be utilised in other industries. Thus by definition, biodiesel is low alkyl esters of long hydrocarbon chain fatty acids prepared from vegetable oils and animal fats through chemical or by biochemical process of transesterification.

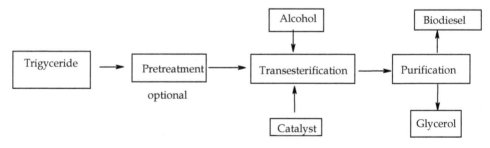

Scheme 2.

2. Feedstock of biodiesel

Different feedstocks have been explored for extraction of vegetable oils in order to transform it to biodiesel. The feedstocks are animal fats, renewable plant resources basically from Euphorbiace family viz. Jatropha caracas, Soya, Sunflower, Castor seeds etc. besides waste cooking oils from different restaurants and food processing industries. Considering several aspects castor oil from castor seeds seem to be an alternative promising feedstock for commercial production of biodiesel particularly for cold climatic regions.

2.1 Why castor oil

Although from the economic point of view waste cooking oils from different sources is a better choice for biodiesel preparation compared to all other sources and vegetable oils, considering the multifarious advantages oil from castor seeds from *Ricinous Communis* (Palma christi)- a species from Euphorbeace family is believed to be a better option. Because castor oil is possibly the plant oil which is industry's most unappreciated asset that contains about 90% ricinolic acid as the major constituent. The plant originates in Africa but now is available in all the tropical and subtropical countries. The plant can stand long periods of drought. The oil has versatile utility such as cosmetics, lubricants, brake fluids, softener in tanning, solar cell, textile company, small components of PC, mobile phones, boots and shoe manufacturing etc. Presently India is the largest producer of castor oil in the world with China and Brazil being the next two. India exports about 15000 tonnes castor seeds per year and 1,00,000 tonnes of castor oil annually to European Union and the domain has been increasing rapidly. In the seed, the oil content is about 50% of the total weight. It is the only unique oil which has an unusual chemical composition of triglyceride of fatty acid. It is the only source of an 18-carbon hydroxylated fatty acid *viz* ricinoleic acid with one double bond. It is reported that fuels having fatty acids with 18 or more carbon atoms and one double bond have viscosity low, higher cetane number and lower cloud and pour point properties are better. From that point of view alkyl ester from castor oil satisfies most of the criteria with the exception of viscosity and cetane number to stand as promising biodiesel candidate. The chemical composition of castor oil triglyceride (castoroil.in - the Home of castor oil in Internet) is

1. Ricinolic acid- 89.5%
2. Linoleic acid- 4.2%
3. Oleic acid- 3%
4. Stearic acid- 1%
5. Palmitic acid- 1%
6. Dihydroxy stearic acid- 0.7%
7. Eicosanoic acid- 0.3%
8. Free fatty acid in refined castor oil- 8.45%

Although considerable researches have been done on palm oil, soya oil, sunflower oil, coconut oil, rapeseed oil, tung oil, jatropha oil etc not much informations are available on castor oil as biodiesel even though it is currently undergoing a phase of active research in several institutions.

Production of castor oil worldwide is 0.5 million tonnes per annum. Consumption of petrodiesel per day is approximately 10 million tonnes. If the entire petrodiesel is to be replaced by castor biodiesel it needs to produce 7000 times the castor oil that is being produced today. However, since it is one of the oldest traded goods mankind has been trading a few thousand years ago, it has a lot of industrial usages and therefore market is already in existences. Further, as the plantation of castor plant has been cultivated commercially, its biology is well understood and high yield hybrid is available. It can also be found in medium climate areas as an annual crop or in tropical area as a small tree. It gives faster oil yield and can be planted as marginal plant in unattended idle areas. The gestation period of harvesting the plant for oil is 4-6 months only.

2.2 Properties of biodiesel from castor oil

The biodiesel prepared from castor oil has certain properties that are attractive particularly for cold climate. It may be mentioned that it has flash point of 190.7⁰C which is much higher than petrodiesel and other vegetable oil biodiesel. The oil is stable at low temperature and makes it an ideal combustible for region of extreme seasonal weather. From cost point of view although 100% biodiesel from castor oil (B_{100}) seems to be expensive its 10% (B_{10}) or 20% (B_{20}) blending with petrodiesel show good flow properties and further lowers the cloud and pour point. Further, due to its ability of displaying as a solvent, sedimentation does not occur which could otherwise potentially obstruct pipes and filters. However,the oil is sensitive to contamination by ferrous salts and rusts particles.Its higher cooling capacity is a key factor in the conservation of engine components. Considering the technical features, castor oil biodiesel is advisable taking into accounts its renewable resources. Because of its biodegradability and lower emissions, it presents a favourable impact on the environment. Moreover, it could be used as a crop substitution program turning it into a factor that promotes growth in many regions affected by several economic problems. Awareness is there in recent times for cultivation of castor plants boosting rural economy by government and private agencies by establishment of transesterification plant with million tonnes capacity per day, trial run using biodiesel from castor oil by Indian Railways, roadways, IOCL, HPLC etc. In addition to it a national mission on biodiesel has been proposed by the government of India with six micro missions to cover different aspects.

3. Transesterification of vegetable oils

Transesterification of vegetable oils has now come a long way for preparation of biodiesel. There are four basic methods for biodiesel production. These are acid catalysed, base catalysed, enzymatic/microbial transesterification and conversion of the oil to its fatty acids and then esterification to have ester as biodiesel.

3.1 Transesterification catalysts

The transesterification reactions require a catalyst in order to obtain a reasonable conversion rate and the nature of the catalyst must conform to the feedstock. Further, the reaction condition and post separation steps are predetermined by the nature of the catalyst.

Generally, transesterification of vegetable oil is done with methanol or ethanol in presence of a base catalyst such as NaOH, KOH, K_2CO_3, NaOMe, NaOEt, NaOPr, NaOBu etc. A minimum content of water and free fatty acid result in the saponification with consequent formation of soap. Presence of large content of water results in hydrolysis of the product formed. Theoretically 3 moles of methanol are required per mole of triglyceride. As the transesterification reaction of triglyceride is a reversible reaction, the excess of methanol shifts the equilibrium towards the direction of ester formation. Freedman et al (Freedman et al, 1984) suggested that 6:1 molar ratio of alcohol to oil is necessary to get the maximum ester yield thus minimising the concentration of tri, di and mono glycerides.

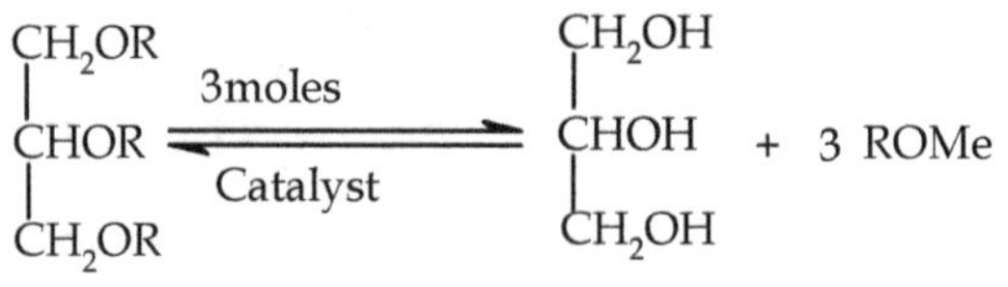

$$\begin{array}{ccc} CH_2OR & & CH_2OH \\ | & \xrightarrow[\text{Catalyst}]{\text{3moles}} & | \\ CHOR & & CHOH \quad + \quad 3\ ROMe \\ | & & | \\ CH_2OR & & CH_2OH \end{array}$$

Scheme 3.

3.2 Solid catalysts for transesterification reactions:
3.2.1 Solid base catalysts
There are reports of many solid base catalysts to be active in transesterification reactions such as supported CaO catalysts (Yan et al, 2008), supported VO_2 catalysts (Kim et al, 2008) various other metal oxides such as BaO, SrO, MgO etc to have transesterified camaelina sativa oil as biodiesel with upto 80-89% yield(Patil and Deng, 2009). However, these solid base catalysts show much lower activity than traditional homogeneous catalysts.Potassium nitrate supported on alumina as solid base catalyst was reported by Vyas et al (Vyas et al,2009) for production of biodiesel from jatropha oil and has been successful in getting 84% yield. Certain of these catalysts are very much sensitive to trace amount of free fatty acid present. Reports of lanthanum based (Kurian et al, 1998) strong basic catalysts have appeared for transesterification and esterification reaction.

3.2.2 Enzyme catalysis
Over the last few decades considerable research have been done on the use of enzyme in transesterification using lipase enzyme from filamentous fungi and recombinant bacteria under various condition. However not considerable attention has been received except in China where 20,000 tonnes of biodiesel per year(Du et al, 2008) is produced. But due to large reaction volume, time, higher conc. of catalyst, cost ($1000 per kg), loosening of catalyst activity on repeated use the process is not commercially viable although friendly to the environment.

3.2.3 Acid catalysts
Homogenous acid catalysts such as H_2SO_4, HCl, sulfonic acid etc. have the potentials to replace base catalysts since they do not show measurable susceptibility to free fatty acid (FFA) and can catalyse esterification and transesterification simultaneously (Kulkarni et al, 2006). However, separation problem, requirement of high temperature, high molar ratio of oil and alcohol, serious environmental and corrosion related problem make their use non practical for biodiesel production.

The demanding feedstock specification for base catalysed reactions have led researchers to seek catalytic process alternative that can ease this difficulty and lower production cost. To eliminate the corrosion, environment problem and time saving for multiple reaction, solid acid catalysts have recently replaced liquid acids for biodiesel production by simultaneous esterification and transesterification. Methodologies based on acid catalysed reaction have the potential to achieve this since acid catalysts did not show measurable susceptibility to FFAs. Compared to homogenous acid catalysts heterogeneous solid acid catalysts have great potential due to advantage in separation and corrosion related problems and such catalysts having large-pores, moderate to strong acid sites and a hydrophobic surface are idea for biodiesel production.

3.2.4 Solid acid catalysts
There have appeared in the literature several solid supported acid catalysts such as heteropolyacid, having Keggin structure viz-12-tungsto-phosphoric acid impregnated on various solid supports like hydrous zirconia (Kulkarni et al. 2006), silica, alumina, and activated carbonate using as solid acid catalyst for biodiesel preparation from different feedstock with achievement of more than 77% yield of biodiesel. Zeolites (Lotero E et al, 2005, Wang et al,2009) with large pore size have been used with success with fatty acid esterification

but at higher temperature. Few other solid supported catalysts for esterification and transesterification of vegetable oils are zeolites with different pore size framework of Si/Al ratio and proton exchange level. These characteristics permit tailoring important catalytic properties such as acid strength. It was observed that zeolite catalysis in transesterification/esterification reaction using large molecules takes place on the external surface of the zeolite catalysts. However, it requires high temperature and the reaction rate is slow. The reactivity on such solid surface catalysts depends upon acid site strength and hydrophobicity of the surface. In fact, pore size, dimensionality of the catalyst channel system related to the diffusion of the reagents and products and aluminium content of zeolite framework strongly affect the zeolite catalytic activity for esterification. Related to zeolites, but with amorphous pore walls, silica molecular sieves such as MCM-41, mesoporous materials are generally not sufficiently acidic to catalyse esterification reaction due to pure silica structure. However introducing aluminium, zirconium, titanium or tin compounds into silica matrix of these solids can significantly improve their acidic properties. However, metal doped materials behave more like weak acids and can only be used for reactions that do not require a strong acid catalyst. It has also been reported that SO_4^{-2}/ZrO_2 has been shown to have applicability for several acid catalysed reactions. However the problem is that SO_4^{-2}/ZrO_2 deactivates in presence of water due to leaching of SO_4^{-2} either in the form of H_2SO_4 or HSO_4^-. Sulphated tin oxide (SO_4^{-2}/SnO_2) prepared from m-stannic acid has shown activity superior to that of SO_4^{-2}/ZnO_2 for esterification of octanoic acid by methanol at 150⁰C due to superior acid strength (Furuta, S. et al 2004). The use of solid catalyst to produce biodiesel requires a better understanding of the factors that govern their reactivity. Thus, an ideal solid catalyst should show some underlying characteristics such as an interconnected system of large pores, moderate to high concentration of high acid sites and a hydrophobic surface. Large interconnected pores would minimise diffusion problem of molecules having long alkyl chain and strong acid sites are needed for the reaction to proceed at an acceptable rate. It is recently attracted considerable attention for solid acid catalyst such as Bronsted acid zeolites, ion exchange resin, metal oxides viz sulphated zirconia WO_3/ZrO_2, MoO_3/ZrO_2, sugar based catalyst (Zong et al, 2007). It has been noted that Bronsted acid catalysts are active mainly in esterification while Lewis acid catalysts are active in transesterification reaction. Therefore, preparation of such solid supported catalysts that contain both Bronsted acid and Lewis acid catalyst site having enhanced water tolerance and large pores, hydrophobic surface and low cost is still a challange. National Chemical Laboratory, Pune India has developed a novel solid double metal composition for transesterification of vegetable oils containing up to 18% FFA to biodiesel (Sree Prasanth et al, 2006). A series of layered alumino silicates with H_2SO_4 impregnation has been reported for transesterification. Activated montmorillonite KSF showed 100% conversion of transesterification within 4 hour at 200⁰C and 52 bar pressure. However problem encountered is leaching, for which reimpregnation of H_2SO_4 on the clay surface is required for reusability. Several other solid acid catalysts were reported but needed higher temperature (>200⁰C) for conversion. The use of age old polymer matrix Amberlyst-15 has also been reported but need mild condition to avoid degradation (Vicente et al,1998).

4. Materials and methods

In view of the above and having observed certain advantages of castor oil over others it was studied the transesterification of it using a simple, cheap and easily prepared solid supported acidic catalyst considering the positives of this clean catalysts.

$KHSO_4$, an acidic salt is ordinarily used as dehydrating agent of alcohol to olefinic compounds at high temperature. It was observed in our earlier study (Goswami et al, 2007) that treatment of it as such with an ester in presence of an alcohol, the ester undergoes partial transesterification very slowly. Dispersing this acidic salt on microporous surface silica gel uniformly triggers transesterification (Goswami et al, 2007) of esters in simple alcohol very satisfactorily giving the product yield more than 95%. The system behaves in a completely different manner on treatment with olefin (Das et al, 2010) leading to dimerization through C-C bond formation or addition product with alcohols depending upon the condition applied. Application of this system to castor oil triglyceride in methanol at its boiling point in 5-6 hours transform it into methyl ester of ricinoleic acid the primary constituent of castor oil along with other fatty acid methyl esters present in it with more than 95% yield.

4.1 Experimental condition
Instruments : The GC was recorded on Chemito 1000 GC using column OVIE+SP2401 (2mX10.635 cm, od) glass column and nitrogen as carrier gas. The textural properties were recorded on Quantachrome Automated Gas Sorption system. The FTIR was recorded on Perkin Elmer System-2000 and FT NMR was recorded on Bruker Avance-DPX-300MHz instrument. Reagents: Castor oil was obtained from local grocery shop (Dabur, 99%). methanol (99.8%) from Fisher Scientific, potassium bisulphate (98%) from Rankem and silica gel (60-100 mesh) were taken from Aldrich Chemicals. Methanol taken was made super dry following standard method:

4.1.1 Catalyst preparation
Potassium bisulphate ($KHSO_4$) 20 gm (144mmol) was dissolved in 100ml distilled water to have a clear saturated solution. The solution was soaked completely in microporous silica (40gm). The soaked mixture was thoroughly mixed and dried in a hot air oven at 150°C for 24 hours to have a free flowing powdery solid. The dried solid mixture was than kept in vacuum desiccator to use as a stock solid supported catalyst (A) in different reactions.

4.1.2 Experimental procedure
25 ml of refined castor oil containing 8.4% FFA was charged with 1 litre dry methanol in a 1.5 litres round-bottomed flask fitted with a condenser and fused calcium chloride guard tube on a preheated oil bath under vigorous stirring. To it was added 1.25gm (5%) catalyst A and stirred at 600 rpm under heating at 70°C (external) for 5 hours. Occasionally TLC was monitored to check the progress of the reaction. After completion, the reaction mixture was distilled to recover methanol. The product with the catalyst remained after separation of methanol was obtained with glycerol as a separate layer. Methyl ester of castor oil along with glycerol layer was decanted out from the solid catalyst surface. Glycerol separated as the bottom layer was taken out from the methyl ester of castor oil (CastMe) layer. The solid catalyst was washed several times with petroleum ether and dried at 150°C for 24 hours in a hot air oven for subsequent runs. The product isolated was found to have yield 95%. During the period of the reactions, samples were taken out at regular intervals and analysed on GC (Fig. 1) using carrier gas nitrogen at flow rate of 2.5kg/cm². Triglyceride, diglyceride, monoglyceride and methyl ester CastMe as transesterified product were quantified by comparing the peak areas of their corresponding standard.

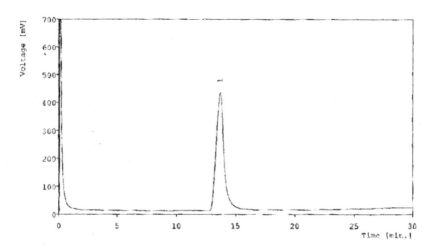

Fig. 1. GC of standard ricinoleic acid methyl ester

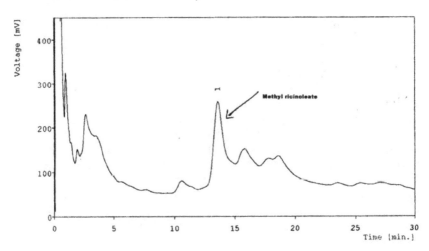

Fig. 2. GC of CastMe

4.1.3 Physical properties of CastMe determined by ASTM D6751 standard

The physical properties of CastMe viz. kinematic viscosity, density, pour point, and cloud point have been determined following standard ASTM D675 method and given in the Table 1 along with reported (Forero, C.L.B., 2004) values of corresponding castor oil, petrodiesel and methyl esters of few other vegetable oils. In Table 2 suggested ASTM standard for pure biodiesel (100%) were given. The properties of CastMe are comparable to those of petrodiesel and acceptable within what is specified for 100 % pure biodiesel as per ASTM standard except that of viscosity and cetane numbers which are the bottlenecks. However, 10% or 20% blended CastMe with petrodiesel that are known as B_{10} and B_{20} have their kinetic viscosity 4.54 & 4.97 mm^2/s and are within ASTM standard.

The corresponding significant FT IR frequencies (Fig 3) and superimposable FTIR spectra of CastMe and standard methyl ricinoleate (Fig 4) and 300 MHz NMR spectral data (Fig5) of methyl ester of castor oil (CastMe) have been as given below.

FT IR (Cm⁻¹, thin film): 1742 (COOMe str),2855 & 2928(CH str),3407(OH str) (Fig 3):

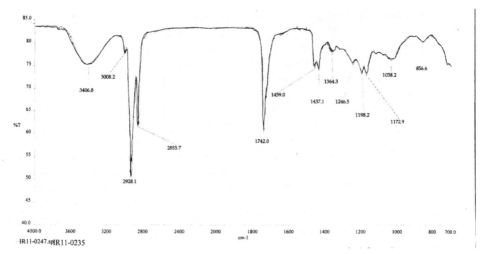

Fig. 3. FTIR of CastMe

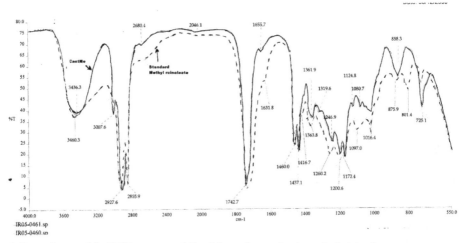

Fig. 4. Superimposable FTIR spectra of CastMe and standard methyl ricinoleate

¹HNMR(δ ppm,CDCl₃) :. 1.96(s,1H,OH), 1.23-2.24(m, nH,(CH₂)ₙMe), 3.55-3.57(m,1H,CH-OH), 3.57(s,3H,COOMe), 5.25-5.47(m, nH, olefinic protons),(n=different numbers of protons of fatty acids), the hydroxyl group present in ricinoleic acid, the major constituent of castor oil imparts unique properties. Because of the branching created by it causes the low cetane number and higher viscosity (Knothe et al, 2008). However the advantage of the present method is that unlike other acidic catalyst, this catalyst system does not facilitate any

methanol olefin etherification (Goodwin et al,2002) even though the constituent of the oil do possess olefinic bonds .

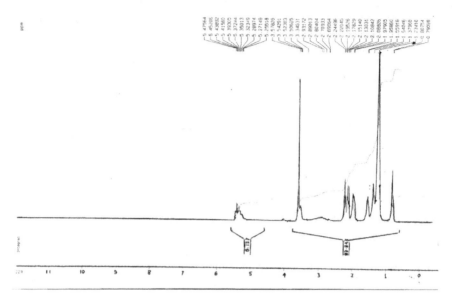

Fig. 5. ¹HNMR (300MHz) of CastMe

Item	Density (g/cc)	Kinematic Viscosity (CST)(38⁰C)	Pour point (⁰C)	Cloud point (OC)	Flash point	Cetane no.
Castor oil	0.963	297	-32	-20	260	
CastMe	0.34	9.4	-45	-23	190.7	42
Petrodiesel	0.86-0.95	3.81	-6	-15	68.3	47
Soy ME	0.885	4.8-4.3	-3.8	-0.5	131	48
Rape ME	0.883	4.53	-10.8	-4.0	170	48
Tallow ME	0.876	51.15	9	13.9	117	35
Canola ME	0.88		-9	1	163	48

Table 1. Physical values of castor oil methyl ester (CastMe) determined along with values of other vegetable oil methyl esters

Property	ASTM standard	limit	units
Flash point	93	100	⁰C
Carbon residue	4530	0.050	wt%
Sulphated ash	874	0.020	wt%
Kinematic viscosity	445	1.9-6.0	mm²/s
Sulphur	2622	0.05	wt%
cetane	613	40	⁰C
Cloud point	2500	By customer	⁰C
Free glycerol	GC	0	wt%

Table 2. Suggested standard for pure (100%) biodiesel as per ASTM.

5. Results and discussion

Potassium bisulphate (PBS) impregnated microporous silica has been evaluated as solid acid catalyst for biodiesel production from refined castor oil containing 8.4% FFA compared to other support viz. alumina with 95% yield. The determination of surface area, pore volume and pore diameter and also FTIR spectra of $KHSO_4$ supported on microporous silica revealed that $KHSO_4$ is well dispersed very evenly generating Bronsted acid site that is responsible for its higher activity.The FT IR spectrum of pure $KHSO_4$,pure silica gel and $KHSO_4$ supported silica gel (Fig-6) have been depicted below.

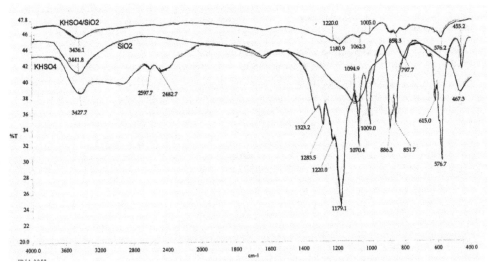

Fig. 6. FT IR spectra of pure $KHSO_4$,Pure SiO_2 and $KHSO_4$ supported on SiO_2

The pure silica FTIR spectra of $KHSO_4$ exhibited typically six major bands located at 577,852,886,1009,1070 and 1179 cm^{-1} which are stretching modes of oxygen bonded to sulphur and hydrogen. In supported $KHSO_4$ catalyst no clear bands were observed. These results indicated that $KHSO_4$ is highly dispersed on the surface of support SiO_2. A 40:1 alcohol to oil ratio at 70⁰C (external) temperature and 5 wt% catalysts loading gave a maximum yield of CastMe up to 95%.

The textural properties (Kulkarni et al, 2006) of the catalyst were summarized in Table 3. The surface area of microporous silica of 60-100 mesh particle size has 300m^2/g and pore volume 1.15cm^2/g and its average pore diameter is 150 A⁰. After loading 50 wt% of $KHSO_4$ the accessible surface area of silica gel left was only 55.45m^2/g and pore volume and average pore diameter were reduced to 0.13cm^2/g and 98.9 A⁰. The reason is attributed to uniform dispersing of KHSO4 on the surface leaving only 55.45m^2/s surface and pore plugging of the support. The same reaction when carried out in a similar fashion supporting $KHSO_4$ on alumina surface, the reaction gives very poor or no yield at all. It may be due to too narrow micropores of alumina which cannot accommodate $KHSO_4$ molecule to disperse uniformly to enhance catalytic activity (Kulkarni et al, 2006) although its surface area is higher (260m^2/g). Even though alumina is an interesting support it is assumed that the surface basicity could bring about decomposition of $KHSO_4$. It means that particles of

KHSO$_4$ conform to silica gel particles in order to disperse on its surface. Large pores can easily accommodate a bulky triglyceride molecule giving KHSO$_4$/SiO$_2$ large active site and surface area resulting in highest activity (Igarashi et al, 1979; Furuta, 2004 and Lecleroq et al, 2001).

Solid support	Surface area (m^2/g)	Pore volume (cm^2/g)	Pore diameter (A^0)
SiO$_2$	300	1.15	150
KHSO$_4$/SiO$_2$	55.45	0.13	98.9

Table 3. The textural properties determined for SiO$_2$ and KHSO$_4$/SiO$_2$

6. Mechanism

The mechanism of the reaction has been shown in Scheme 4. The interaction of the carbonyl oxygen of the ester with the conjugate acid potassium ion from the silica surface of the catalyst forms carbocation by enolizing it. The carbocation is stabilized by the bisulphate ion and facilitates nucleophilic attack methanol on the carbocation producing a tetrahedral intermediate (c).

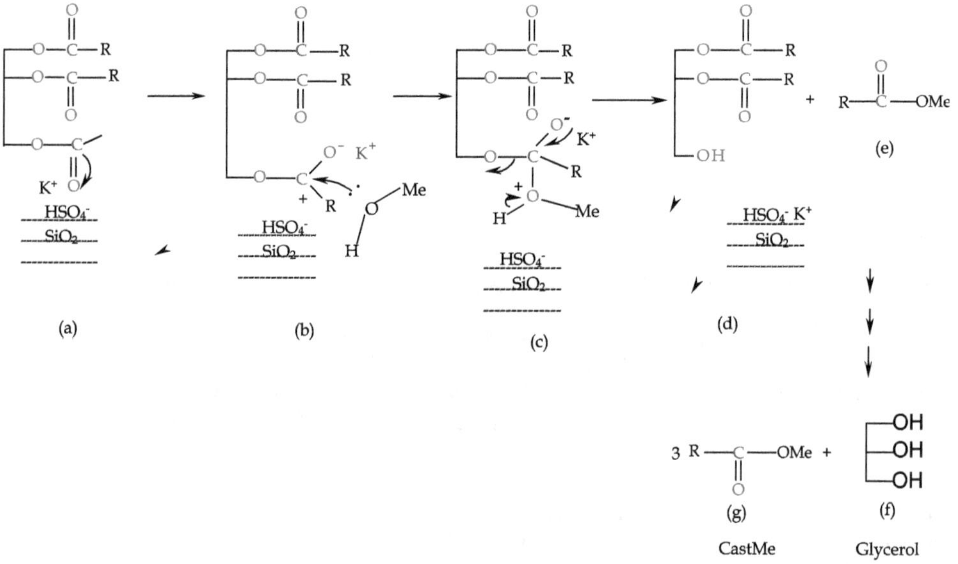

Scheme 4.

In the reaction sequence the triglyceride was converted stepwise to di and mono glyceride and finally to glycerol. The tetrahedral intermediate (c) formed during the reaction eliminate di, monoglyceride and glycerol when tri, di and monoglyceride came in contact with the acidic site respectively to give one mole of ester in each step. It has been reported (Freedman, B, 1986) in fact that the rate limiting step varied over time and in three stages in accordance with the observed reaction rate could categorize the overall reaction progress. In the first stage the reaction was characterized by a mass transfer controlled phase in which the low miscibility of the catalyst and the reagent or the non-polar oil was separated from the polar alcohol phase. The second phase is product formation stage whereby the product formed acts as an emulsifier. It is a kinetically controlled stage and is characterised by abrupt range of product formation. Finally the equilibrium is reached at the completion stage. It was found in castor oil transesterification with 40:1 alcohol to oil ratio acceptable reaction rate was achieved. Thus from this observation it can be stated that the forward reaction is pseudo first order kinetics while the backward or the reverse reaction is second order kinetics.

7. Influence of reaction parameters

The transesterification of castor oil in presence of $KHSO_4$ supported on silica gel in methanol is influenced by certain reaction parameters which have been studied thoroughly varying the conditions at different stages and the results have been appended below..

7.1 Reaction temperature
Initially the transesterification reaction was attempted at room temperature under stirring at 600 rpm for more than 48 hours. However the reaction rate at room temperature was found to be very slow and only 30-35% conversion was observed. It means that the rate of reaction is influenced by the reaction temperature. Gradually when the reaction temperature was raised by 10^0C the reaction rate is increased with increase of product formation and at 70^0C (external) temperature the formation of the product was found to be maximum of 95%. Beyond this temperature there was found to be no further increase of yield (Fig. 7).

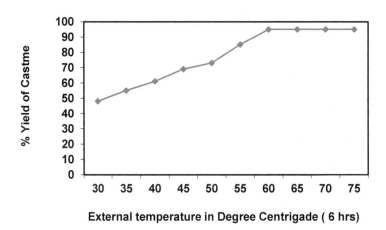

Fig. 7. Effect of external temperature on the reaction course

7.2 Effect of time
The effect of reaction time was studied and result was shown in Fig. 8. It was found that increasing the reaction time upto 5 hours enhanced the castor oil methyl ester yield and

beyond it there found to be no further improvement. It means 5 hours time is optimum period required.

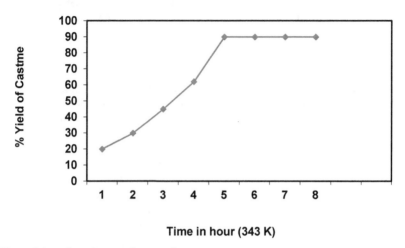

Fig. 8. Effect of time duration on the reaction course

7.3 Effect of alcohol to castor oil ratio
Methanol to castor oil weight ratio is one of the important parameters that affect the yield of methyl ester of castor oil. Theoretically the transesterification reaction requires 3 moles of methanol per mole of triglyceride (Lotero et al, 2005). Since the reaction is a reversible one, the excess methanol shifts the equilibrium towards the direction of ester formation (Cannkei et al, 1999). Generally heterogeneous acid catalytic of transesterification reaction is well known for slow reaction rate. In order to improve the rate of this reaction, use of excess alcohol is an option. It was reported (Xie et al, 2005) that increase of the ratio up to 275:1 of alcohol to oil improves the rate of transesterification reaction.In the present work with preoptomized reaction parameters the methanol to castor oil ratio was varied in the range 5:1 to 40:1 and its influence on the yield of CastMe was investigated at the end of 5 hours. It was clearly observed that at 70⁰C (external) temperature with increase in ratio of alcohol to oil from 5:1 to 40:1 increased the yield of CastMe from 75% to 95%. Presence of 8.4% FFA in the refined castor oil did not affect the activity of the catalyst. Further increase of methanol did not show any significant improvement (Fig 9). The excess methanol can be recovered for reuse and low cost of methanol makes it the first choice for transesterification.

7.4 Effect of catalyst amount
The catalyst amount is also an important parameter that needs to be optimized for increasing the yield of castor oil methyl ester(CastMe). The effect of $KHSO_4/SiO_2$ wt/wt of castor oil on the reaction was studied. At low catalyst amount (< 5 wt %) there were not enough active site for reaction. The optimum amount of catalyst employed was found to be 5 wt% of castor oil to isolate a yield of 95% of the product (Fig. 10).

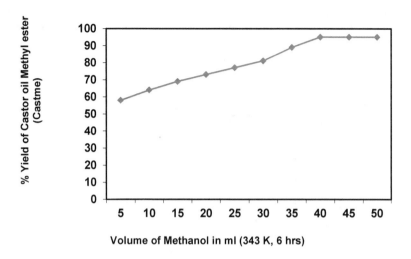

Fig. 9. Effect of volume of alcohol on the reaction course

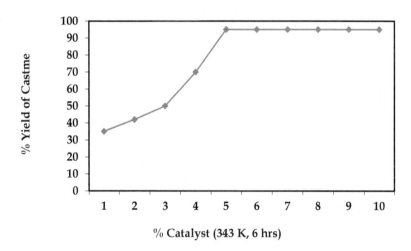

Fig. 10. Effect of catalyst amount

7.5 Catalyst recycling

The cost of a process depends upon the recyclability of a catalyst. It has been found that the dispersed catalyst $KHSO_4$ on silica gel surface after the transesterification reaction of castor oil in methanol, a certain amount gets leached out with methanol either in the form of H_2SO_4 or in HSO_4^-. However, after the completion of the reaction, methanol is distilled out completely and methyl ester of castor oil (CastMe) was extracted in dichloromethane whereby KHSO$_4$ is retained on the surface of silica. The catalyst was washed several times with petroleum ether and then dried completely at 150ºC for 8-10 hours. On use of this catalyst for 5 runs with same amount of castor oil and methanol the yield of CastMe decreased was subtle even at fifth reuse (Fig. 11).

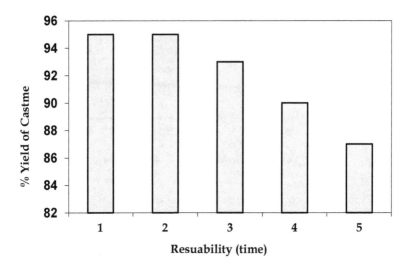

Fig. 11. Catalyst recycling

8. Conclusion

Silica gel supported KHSO$_4$ acidic catalyst prepared for production of biodiesel from refined castor oil containing 8.4% free fatty acid has been found to be a simple, cheap, ecofriendly and recyclable catalyst system for excellent yield of castor oil biodiesel under mild condition. The activity of the catalyst system is not affected by the presence of free fatty acid. The system is so simple that it does not require any special design compared to other solid supported acidic catalysts.

It may be mentioned in this context that the leading oil companies in the whole world are looking to tap the business opportunities of biodiesel. In the developed process such as the one discussed in this chapter is scaled up to commercial levels by more and more oil companies, it could be a major step towards creation of an eco-friendly transportation fuel that is relatively clean on combustion and provides farmers with substantial income.

9. References

Akoh, C.C., Chang, S.W., Lee, G.C. and Shaw, J.F. (2007). Enzymatic approach to biodiesel production. *J. Agric. Food. Chem.*, 55, 8995-9005. DOI: 10, 1021/jf071724y

Canakei,M. and Van,G.J. (1999). Biodiesel production via acid catalysis *Trans. ASAE*, 49,1203-1210.

http:/www.castoroil.in-The Home of Castor oil in the Internet. (2010). Castor oil chemicals and derivatives: Chapter3 ,*Comprehensive Castor Oil Report:*,pp 10

Das, R.N., Sarma, K. and Goswami, A. (2010). Silica supported potassium bisulphate: an efficient system for activation of aromatic terminal olefins, *Synlett*,19,2908-2912

Du, W., Li,W., Sun,T.,Chen,X. And Liu, D. (2008). Perspectives for biotechnological production of biodiesel and impacts, *Applied Microbiol.Biotechnol*, 79,331-337, DOI: 10.1007/c00253-008-1448-8

Forero, C.L.B. (2004). Biodiesel from castor oil : a promising fuel for cold weather, *Report from Francisco de Paula Santander University*, Columbia, www.icrepq.com

Freedman, B; Butterfield, R.O. Pryde B.H.(1986). Transesterification kinetics of soyabean oil,*J. Am. Oil. Chem. Soc.* 63, 1375-1380.

Freedman, B., Pryde, E.H and Mounts, T.L. (1984). Variables affecting the yields of fatty esters from transesterifies vagatable oils. *J.Oil.Chem Society*. 61, 1638-1643, DOI: 10.1007/B IO 2541649, 2008.

Furuta, S, Matsuhashi,H, Arata K. (2004). Catalytic action of sulphated tin oxide for etherification and esterification in comparison with sulphated zirconia. *Appl-Catal A*. 269, 187-191.

Furuta, S., Matsuhashi ,H. and Arata ,K. (2004).Performance evaluation of biodiesel production by solid/liquid catalysts, *Catalysis Commun.*, 5,5721-5723

Ghaly, A.E., Dave , D., Brooks,M.S. and Budge, S. (2010). Production of biodiesel by Enzymatic transesterification: Review. *American Journal of Biochem. and Biotech* 6(2) 54-76.

Goodain, J.G.; Natesalehawat. S; Nikolopolous A.A Kim. S.Y. (2002). Etherification on zeolites: MTBE Synthesis, *Catal., Rev. Sci. Eng.* 44,287-320

Goswami,A., Das,R.N. and Borthakur,N. (2007). An efficient selective solid supported system for deprotection of alcohols from esters, *Ind,J Chem*,46B,1893-1895

Igarashi, S., Matsuda, T and Ogino, Y. (1979). Catalysis for vapour phase synthesis of methyl tertiary butyl ether,*J. Jpn Petrol.Ins.* 22, No, 331-336.

Kim, M., Yan,S.,Wang,H., Wilson, J.,Salley,S.O., Simon Ng, K.Y. (2008). Catalytic activities of vanadia loaded heterogenous catalysts on transesterification of soyabean oil with methanol, *99th AOCS Annual Meeting* in Seattle, Washington

Knothe, G. (2008). 'Designer' Biodiesel, optimozing fatty ester composition to improve fuel properties, *Energy & Fuels*, 22, 1358-1364.

Kulkarni, M.G, Gopinath, R., Meher, L.C. and Dalai A,K. (2006). Solid acid catalysed biodiesel production by simultaneous esterification and transesterification, *Green Chem*, 1056-1062.

Kurian J.V., Liang,Y. (1998). Transesterification process using lanthanum compound catalyst, *US Patent* 5840957

Lecleroq, E. Finiels, A. and Moreau,C.(2001). Transesterification of Sunflower oil over zeolite using different metal loading: A case of leaching and agglomeration studies,*J. Am Oil Chem Soc.*,11,1161.

Lotero, E., Liu,Y., Lopez, D.E., Suwannakarn K, Bruce, D. A and Goodwin, Jr. J.G. (2005). Synthesis of biodiesel via acid catalysis, *Ind.Eng. Chem. Rev*, 44, 5353-5363.

Lou, W.Y. Zong,M.H. and Duan Z.Q. (2008). Efficient production of biodiesel from high free fatty acid containing waste oils using various carbohydrate-derived solid acid catalysts, *Bioresource Technology*, 99(18), 8752-8758.

Patil, P.D. and Deng, S. (2009). Transesterification of camaelina sativa oil using heterogenous metal oxide catalysts, *Energy Fuels*,23,4619-4624

Robels-Medina, A., Gonzalez-Moreno,P.A., Esteban-Cerdan, L.   and Molina-Grima, E. (2009). *Biocatalysis Towards Ever Greener Biodiesel Production. Biotechnol Adv.*, 27, 398-408. DOI: 100.1016/j, biotechadv, 2008, 10.008.

Singh, A.K. and Fernando, S.D. (2008). Transesterification of soyabean oil using heterogenous catalysts, *Energy & Fuels*, 22,2067-2069

Sreeprasanth, P.S., Srivastava, R., Srinivas, D and Ratnaswami P. (2006). Hydrophobic solid acid catalysts for production of biofuel and bolubricantst *Appl. Catal A. General* 314, 148-159.

Vicente,G., Coteron, A., Martinez, M., Aracil, J. (1998). Application of the factorial design of experiments and response surface methodology to optimize biodiesel production, *Ind. Crop Products*, 8,29-35

Vyas, A.P., Subramanyam, N. and Patel, P.A. (2009). Production of biodiesel through transesterification of jatropha oil using potassium nitrate supported on alumina solid catalyst, *Fuel*, 88,4,625-628

Xie, W; Peng, H and Chen, L. (2006). Calcined Mg-Al hydrotalcites as solid base catalysts for methanolysis of soyabean oil *J. Mol. Catal A: Chem*, 246, 24-32.

Xu, G. And Wu, G.Y. (2003). The investigation of blending properties of biodiesel and no. O diesel fuel. J. *Jiangsu Polytechnique University*, 15; 16-18.

Wang, J.,Chem, Y, Wang, X and CaO, F. (2009). Aluminium dodecatungstophosphate ($Al_{0.9}H_{0.3}PW_{12}O_{40}$) nanotube as a solid acid catalyst one.pot production of biodiesel from waste cooking oil, *BioResources* ,4(11), 1477-1486.

Yan,S., Lu,H. and Liang, B.(2008). Supported calcium oxide catalysts used in the transesterification of rapeseed oil for the purpose of biodiesel production, *Energy and Fuel*, 22(1),646-651

Zong, M.H., Duan Z.Q., Lou, W.Y., Smith, T.J. and Wu, H. (2007). Preparation of a sugar catalyst and its use for highly efficient production of biodiesel, *Green Chem.* 9(5), 434-437.

7

Biodiesel Production with Solid Catalysts

Feng Guo and Zhen Fang

Chinese Academy of Sciences, Biomass Group,
Xishuangbanna Tropical Botanical Garden,
China

1. Introduction

Biodiesel is usually produced by transesterification of vegetable oils or animal fats with chemical catalysts, especially in the presence of strong acidic or basic solutions, such as hydrochloric acid, sulphuric acid, sodium hydroxide, sodium methoxide and potassium hydroxide. Homogeneous alkali catalysts can convert triglycerides to their corresponding fatty acid methyl esters (FAMEs) with high yield, less time and low cost. However, separating the catalyst from the product mixture for recycling is technically difficult. After reaction, the catalyst should be neutralized or removed with a large amount of hot water, which will produce a large amount of industrial wastewater.

Typical plant oils, such as soybean oil, rapeseed oil and palm oil, are the main edible oils. They are not suitable as raw materials, particularly in developing countries due to limited supply and high cost. Therefore, low-cost lipids, such as non-edible oils (e.g., Jatropha oil), animal fats and waste oils, are used as ideal feedstocks. Such oils usually contain some water and free fatty acids (FFAs) that will form soap when homogeneous base catalysts are used. On the other hand, homogenous acid catalysts are corrosive to equipment. Solid heterogeneous catalysts are used to overcome these problems, because they are non-corrosive, non-toxic, and easily-separated for recycling. Reusability of heterogeneous catalysts makes continuous fixed-bed operation possible. Such continuous process can minimize product separation and purification costs, make it economically viable to compete with commercial petroleum-based diesel fuel. This chapter describes solid heterogeneous catalysts for biodiesel production and their typical catalytic mechanism.

2. Heterogeneous solid catalysts

In laboratory-scale experiment, heterogeneous processes could be run in a continuous mode with a packed-bed continuous flow reactor. Heterogeneous catalysts were easily separated from the products, water-washing process and neutralization steps were avoided. Contaminated water from this process was greatly reduced, and the sewage treatment fees were also minimized. New types of heterogeneous catalysts have mushroomed and developed in recent years.

2.1 Heterogeneous acid catalysts

Acid catalysts can simultaneously catalyze both esterification and transesterification, showing a much higher tolerance to FFAs and water than basic homogeneous catalysts (e.g.,

NaOH and KOH). Homogeneous acid-catalyzed reaction is about 4000 times slower than the homogeneous base-catalyzed reaction (Lotero et al., 2005). Heterogeneous acid catalysts performed less activety, but they are favorable for low-qualified oil feedstocks with high FFAs. Now, synthetic solid acids have already amounted to hundreds of species, most of them can be used in esterification and transesterification reactions. Solid acids keep stable activity in conversion of low-qualified oils or fats to biodiesel. Currently developed solid acid catalysts are introduced in the following sections: cation exchange resin (i.e. Amberlyst-15 and NR50), mineral salts (i.e. ferric sulfate, zirconium sulfate, alum phosphate and zirconium tungsten), supported solid acid and heteropolyacid catalysts.

2.1.1 Ion-exchange resins

Ion-exchange resins are widely used in important industrial processes for both separation and reaction applications. They are less expensive than lipase and supercritical methanol. Ion exchange resins also help to separate biodiesel and glycerol. Shibasaki-Kitakawa et al. (2007) found catalytic activity of anion-exchange resins correlated positively with cross-linking degree and particle size. The activity of acid ion-exchange resins for the esterification reaction is influenced by the accessibility of reactants to the matrix anchored sulfuric acid groups located at the surface or inside the resins. Mass-transfer restriction is another factor affecting catalytic activity. Internal diffusion was found to cause mass-transfer restriction and is rate-limiting for regular resins. Most of the active sites are embedded in the gel matrix, so the resins with macro-pores have high catalytic activity. Furthermore, the catalytic activity decreased when the cross-linking degree of polymeric matrix increased. Reusability is an important evaluation index for industrial applications of resins. Mechanical strength and thermo-stability are important for the large-scale applications of resins in biodiesel production. Ion exchange resins usually don't change catalytic property for long time operation at low temperature (< 100 ºC).

Caetano et al. (2009) studied esterification of palmitic acid with methanol using poly(vinyl alcohol) cross-linked with sulfosuccinic acid (SSA) resin at 60°C, about 90% conversion rate was achieved after 2 h. Only about 5% sulfosuccinc acid was leached after 7 recycles. Activity of NKC-9 resin even slightly increased at the first 10 runs, due to breakdown of resin particles under mechanical agitation (Feng et al., 2010). Continuous production of biodiesel in a fixed-bed reactor packed with resins was successively operated (Shibasaki-Kitakawa et al., 2007; Liu et al., 2009; Feng et al., 2011). After 500 h, conversion yield of FFAs still kept over 98%. Amberlyst-15 performed high activity at 100 ºC in the fixed-bed, and a 97.5% FAMEs yield was achieved (Son et al. 2011). Combination of fixed-bed reactor with supercritical CO_2 may develop a continuous process that is preferred for massive biodiesel production. Catalyst deactivation is caused by salt contaminants and water-swelling. Catalytic active sites on acidic resins can exchange with salt ions contained in oil. Traces of Na, K, Mg and Ca lead to a continuous activity loss (Russbueldt and Hoelderich, 2009). Deactivated catalyst can be recovered to its original activity by acid washing. It was found that temperature has negligible effect on water-swelling, but the water absorbed on the resin surface can be extracted by excessive methanol (Tesser et al., 2010)..

2.1.2 Zeolites

Zeolites are crystalline alumino-silicates with a three-dimensional porous structure. They can be synthesized with different crystal structures, definitive pore sizes, framework Si/Al ratios and adjustable acid centers to have some important catalytic properties. Aluminum atoms and ions in skeletons and porosity supply the acidic sites. Zeolites have extremely high internal surface area (600 m^2/s) and high thermal-stability (1000 °C), as the most popular solid catalysts. Acidic-shape selectivity is a significant feature of zeolite, derived from the influence of pore size and shape on a reaction. Zeolites were used in biodiesel production as a heterogeneous catalyst. Pérez-Pariente et al. (2003) studied the selective synthesis of fatty monoglycerides with Zeolites. Compared with reaction parameters, catalyst properties have more effects on monoglyceride yield.

Zeolite β is a high silica zeolite with both Lewis-acid sites and Brønsted-acid sites, containing an intersecting three-dimensional structure of 12-membered ring channels (Shu et al., 2007). Lewis-acid sites are mainly present in the micro-porous walls. On the contrast, Brønsted-acid sites are present on the internal and external surface. Zeolite β does not exhibit high activity in transesterificaiton, but it can be used for selective removal of FFAs in waste oil (Chung et al., 2008).

HY zeolite has a large number of weak acid sites. When one Si^{4+} is substituted by an Al^{3+}, the zeolite framework generates one Brønsted acid site. On the other hand, one Na^+ cation neutralizes one acid site. Furthermore, hydroxyl groups formed by ion exchange of HY zeolite with ions, such as Ca^{2+}, Mg^{2+} and La^{3+}, can strengthen the acidic sites. Acidity of zeolite can also be adjusted by introducing protons with dilute hydrochloric acid. Sasidharan and Kumar (2004) found that large-pore zeolites such as Y, mordenite, and β showed higher activity (biodiesel yield 92%) than the medium-pore ZSM-5 and aluminum containing mesoporous MCM-41 (biodiesel yield < 30%). The high pore volume of large-pore zeolites favored reaction by rendering the active sites more accessible to the bulky triglyceride molecules. However, Hβ-Zeolite catalyzed transesterification of crude *Pongamia pinnata* oil gave low yield of 59% at a long reaction time (24 h) (Karmee and Chadha, 2005). Internal diffusion resistances are considered to limit reaction rate significantly. Thus, large-pore zeolites are active for the reaction with satisfactory reaction rate.

Most of zeolites exhibit not only acidic property, but they also provide high activity and selectivity in various acid catalysts as carrier. Bifunctional catalyst can usually be prepared by combining active catalytic sites on an acid zeolite. Shu et al. (2007) introduced La ion into zeolite β with $La(NO_3)_3$ as the ion exchange precursor. La/zeolite β resulted in higher conversion with higher stability than zeolite β because it has more external Brønsted acid sites available for the reactants. Triglyceride conversion yield of 48.9 wt% was obtained at 60 °C for reaction time of 4 h.

2.1.3 Heteropoly acids (HPAs)

A heteropoly acid is a class of acid made up of a particular combination of hydrogen and oxygen with certain metals (i.e., tungsten, molybdenum and vanadium) and non-metals (i.e., silicon, phosphorus). Heteropoly acid is frequently used as a re-usable acid catalyst in chemical reactions, but their long term stability and performances are not yet fully characterized. With inherent advantages of strong Brønsted acidity, stability and high proton mobility, HPAs are favorable as environmentally benign and economical solid catalysts. Owing to their unique physicochemical properties, HPAs are profitably used in homogeneous, biphasic and heterogeneous systems. There are many types of heteropolyacids, and the Keggin ($H_nXM_{12}O_{40}$) and Dawson ($H_nX_2M_{18}O_{62}$) structures are two of the better known groups (Kozhevnikov, 1998). HPAs (e.g., $H_3PW_{12}O_{40}$) are soluble in water and possess acidic strength as strong as sulfuric acid. HPAs solubility can be changed via alkali-exchange, and modified HPAs exhibiting significantly higher activity.

HPAs are excellent and environmentally benign acid catalyst for the production of biodiesel, which are tolerant to contaminations contained in oil resources such as FFAs and water. The Keggin HPA (i.e., $H_3PW_{12}O_{40}$) is soluble in methanol, and the use of Keggin heteropolyacids for triglyceride (trans)esterification has been reported. Alsalme et al. (2008) studied the intrinsic catalytic activity of Keggin HPAs, indicating activity of HPAs is significantly higher than that of the conventional acid catalysts in (trans)esterificaiton. Their acid strength in the descending order is as follows: $H_3PW_{12}O_{40}$ > $Cs_{2.5}H_{0.5}PW_{12}O_{40}$ > $H_4SiW_{12}O_{40}$ > 15%$H_3PW_{12}O_{40}$/Nb_2O_5, 15%$H_3PW_{12}O_{40}$/ZrO_2, 15%$H_3PW_{12}O_{40}$/TiO_2 > H_2SO_4 > HY, H-Beta > Amberlyst-15. HPA is able to efficiently promote the esterification with a similar performance to sulfuric acid. However, the recovery and reutilization of HPAs is difficult.

The main disadvantage of HPAs is their solubility in water and polar solvents. This problem can be overcome by converting it into its salt (e.g., ammonium salt) with decreases of acidity and catalytic activity. It is reported that partial exchange of ammonium salt in 12-tungstophosphoric acid with offers more acidic strength to the catalyst than the fully exchanged ammonium salt (Giri et al., 2005). Exchange of protons in HPA can help promote its activity in transesterification of triglycerides. The protons replacement has similar effects on activity as cations concentration increase. $Cs_xH_{3-x}PW_{12}O_{40}$ (x = 0.9-3), one kind of insoluble Keggin HPAs, offers excellent performance in (trans)esterification (Narasimharao et al., 2007). The catalytic activity of Cs-salts decreases as the content of Cs in HPW grows, due to the decrease of pH and the increase of conductivity of colloidal solutions in direct relation with the acidity of surface layers of primary particles. Furthermore, low-Cs loading on HPAs shows some dissolution of an active acid component after reflux in hot methanol, while high-Cs loading on HPAs is stable in hot methanol.

Immobilization of HPAs on carrier is also an efficiency method to obtain insoluble catalyst. Such supported solid acids performed high thermal-stability even under reaction conditions of 200 ºC. Caetano et al. (2008) used tungstophosphoric acid, molibdophosphoric acid and tungstosilicic acid immobilized by sol-gel technique on silica to catalyze esterificaiton of palmitic acid with methanol. The higher heteropolyacids load on silica, the lower the catalytic activity is observed. Tungstophosphoric acid-silica (with 4.2 wt.%) showed the highest catalytic activity, 100% palmitic acid conversion was achieved after 30 h reaction time with methanol. Zięba et al. (2010) tested catalytic performance of Amberlyst-15, Nafion-SAC-13, polyaniline-sulfate, silver and cesium salts of HPAs in transesterifiaction of triglycerides with methanol. $Cs_2HPW_{12}O_{40}$ was the most active catalyst due to its highest strength of acid sites, but the great affinity toward glycerol led to its deactivation during recycling process.

2.1.4 Supported acid catalysts

Supports can provide higher surface area through the existence of pores where acidic sites can be anchored. Supports should be modified during preparation of catalysts to anchor catalytic species and obtain reusability. Furthermore, some amorphous carriers also showed good activity for (trans)esterification. Metal oxides are widely used as catalyst supports because of their thermal and mechanical stability, high specific surface area, and large pore size and pore volume. Because solid acids function the same as H^+ in sulfuric acid for (trans)esterification, sulphonated metal oxides, such as SO_4^{2-}/Al_2O_3, SO_4^{2-}/TiO_2, SO_4^{2-}/ZrO_2, SO_4^{2-}/SnO_2 and SO_4^{2-}/V_2O_5 (Garcia et al., 2008) can supply more acid species. Such solid acids are usually prepared by impregnating the hydroxides from ammonia precipitation of corresponding metal salt solutions with aqueous sulfuric acids followed by calcination. In addition to acid amount and acid strength adjustment, the catalysts are satisfactorily active in a heterogeneous liquid–solid system and are recoverable and reusable.

A cheap and high efficiency solid acid catalyst (SAC) derived from sulfonation of carbonized D-glucose or sucrose was reported, and used in transesterification of vegetable oil with alcohol (Shu et al., 2009; Toda et al., 2005; Zong et al., 2007). The catalyst was prepared from carbohydrates by carbonizing at 400 °C under N_2 atmosphere and then sulphonating at 150 °C. The solid acid catalyst can also be prepared by direct sulphonation of lignin consisting of polyethers and C-C linked phenylpropanes as shown in Fig. 1. The carbon carriers are amorphous, polycyclic aromatic carbon sheets containing SO_3H groups as active sites (Shu et al., 2009; Toda et al., 2005). The polycyclic carbon sheets can absorb long-chain hydrocarbon for reactants in solution to access SO_3H groups. Hydrolysis of cellulose to saccharides using such amorphous carbon bearing SO_3H, COOH, and OH function was studied (Suganuma et al., 2008). Phenolic OH groups bonded to the grapheme sheets can absorb β-1,4 glycosidic bonds and provide good access of reactants in solution to the SO_3H groups in the carbon material.

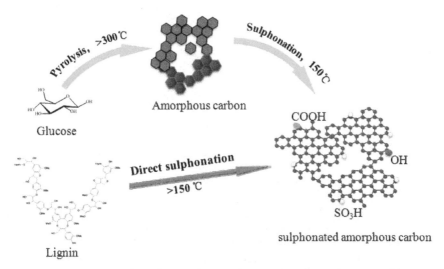

Fig. 1. Preparation process of sulphonated amorphous carbon from glucose and lignin.

Zong et al. (2007) utilized SAC as a solid acid catalyst for transesterification of waste oil (27.8% FFAs) with methanol. The reaction was carried out at 80 °C for 15 h, a high yield of above 90% obtained as compared with below 80% yield when sulfated ZrO_2, Amberlyst-15 and niobic acid were used. SAC was also used for other types of organics reactions, such as oxidations of organic compounds (e.g., sulfides, tertiary amines, aldehydes) with hydrogen peroxide (Shokrolahi et al., 2008). Specific surface area, pore size, pore volume and active site concentration on the surface of catalyst are effective factors on catalytic activity. Stability of the active sites is important for their industrial applications.

2.2 Heterogeneous base catalysts
The transesterification of vegetable oils or animal fats to biodiesel by chemical catalysts, especially in the presence of a strong basic solution, such as sodium hydroxide and

potassium hydroxide, has been widely used in industrial production of biodiesel. Such basic solutions can transform triglycerides to their corresponding FAMEs with higher yield at lower temperature and shorter time than those by acid catalysts. However, separating the catalysts from products is technically difficult. Moreover, natural vegetable oils and animal fats usually contain small amounts of FFAs and water, which can have significant negative effects on the transesterification of glycerides with alcohols, and also hinder the separation of FAMEs and glycerol due to saponification of FFAs. Compared with basic solutions, solid base catalyst is preferred due to easy separation.

Heterogeneous base catalysis has a shorter history than that of heterogeneous acid catalysis. Solid bases refer mainly to solids with Brønsted basic and Lewis basic activity centers, that can supply electrons (or accept protons) for (or from) reactants. Heterogeneous base-catalyzed transesterification for biodiesel synthesis has been studied intensively over the last decade. Low-qualified oil or fat with FFAs and water can be used. However, the catalytic efficiency of conventional heterogeneous base catalysts is relative low and needs to be improved. Various types of catalytic materials have been studied to improve the transesterification of glycerides. Heterogeneous base catalysts, such as hydrotalcites, metal oxides, metallic salt, supported base catalyst and zeolites are introduced herein details.

2.2.1 Hydrotalcites

Hydrotalcites (HTs) are a class of anionic and basic clays known as layered double hydroxides (LDHs) with the formula $Mg_6Al_2(OH)_{16}CO_3 \cdot 4H_2O$. HTs consist of positively charged brucite-like layers and interstitial layers formed by CO_3^{2-} anions, and water molecules compensate the positive charge resulting from the substitution. LDHs have strong alkali sites and high stability with good adjustability of composition and structure. However, low surface area affected its catalytic activity. Mg/Al mole ratio and calcination temperature are the determining factors for the base-catalyzed activities. HTs with a 3:1 molar ratio of Mg to Al have the highest basicity and activity (Xie et al., 2006). Decomposition of HTs after calcination yields a high surface area Mg-Al mixed oxide, which presumably exposes strong Lewis basic sites. During calcination process, the interlayer water is lost first, followed by dehydroxylation and decomposition of interlayer carbonate to CO_2, which generate a porous structure and specific surface area ranging from 150 to 300 m^2/g (Lee et al., 2009). Furthermore, Mg^{2+} can be replaced by Zn^{2+}, Fe^{3+}, Co^{2+}, Ni^{2+}, and Al^{3+} by Cr^{3+}, Ga^{3+}, Fe^{3+}. HTs substituted with copper have a relatively uniform porous structure with decreased specific surface area. For iron substituted HTs, microporosity features developed, pore size decreased and specific surface area increased. The initial study by Cantrell et al. (2005) on biodiesel synthesis with HTs indicated magnesium content has obvious effects on catalytic activity. Increase of both magnesium content and electron-density enhances alkaline of HTs and finally increases biodiesel yield.

As classical solid base materials, calcined HTs were widely used as catalyst in the production of biodiesel (Brito et al., 2009; Deng et al., 2010). The basicity and surface area of HTs can be tuned by modifying chemical composition and preparation procedure. A co-precipitation method usually used to synthesize HTs with Mg/Al molar ratio of 3/1 using urea as precipitating agent. In previous work (Xie et al., 2006), transesterification process was carried out with reflux of methanol, methanol/soybean-oil molar ratio of 15/1, reaction time of 9 h and catalyst amount of 7.5%, and oil conversion rate was only 67%. In the work of Brito et al. (2009), waste oil as feedstock, biodiesel production was performed at temperatures ranging from 80 to 160 °C, methanol/oil molar ratio from 12/1 to 48/1 and catalyst concentration from 3 to 12%, respectively, and 90% biodiesel yield was achieved.

It is known that HTs present lamellar structure thus not pose accessibility restrictions of vegetable oil molecules to catalyst sites. Improvement of specific surface area becomes necessary to obtain high catalytic activity. Deng et al. (2011) synthesized a series of nanosized HTs by a modified co-precipitation method. SEM images of HTs and calcined HTs were given in Fig. 2. Variables of temperature, solution pH and ageing time have a strong influence on the final basicity of the mixed oxides. Mg-Al ratio in the precursor HT depends on the basic properties of these sites. In the transesterification experiment using Mg-Al HT catalysts, 95% biodiesel yield was achieved from *Jatropha* oil in 1.5 h. Pre-mixture of HTs with methanol is essential to optimize catalyst activity to avoid lagging in reaction activity due to mass diffusion. Xi et al. (2008) tested influence of water on the activity and stability of activated Mg-Al HTs. In the presence of certain amount of interlayer water, Brønsted base sites were active. However, high degree of hydration caused rapid deactivation of the catalyst. Mg-Al HT shows relatively robust activity in the presence of water or FFAs tolerate, which is an attractive feature for biodiesel production.

Fig. 2. SEM images of (A) hydrotalcite and (B) calcined hydrotalcite at 500 °C for 6 h.

2.2.2 Metal oxides

Metal oxides are composed of cations possessing Lewis acid and anions with Brønsted base. Metal oxides used in transesterification are classified as single metal oxides (e.g., MgO, CaO and SrO) and mixed metal oxides [A-B-O type metal oxides, where A is an alkaline-earth metal (Ca, Ba, Mg), alkaline metal (Li), or rare earth metal (La) and B is a transition metal (Ti, Mn, Fe, Zr, Ce)] (Kawashima et al., 2008; Liu et al., 2007; Liu et al., 2008; Montero et al., 2009). Early studies on heterogeneously catalyzed transesterification were focused on the catalysis by single metal oxides. The basicity of oxides (especially, basic sites) directly depends on reaction rate. A comparison of several metal oxides (MgO, CeO_2, La_2O_3 and ZnO) indicated that the most basic one is La_2O_3, followed by MgO, CeO_2 and ZnO (Bancquart et al., 2001).

The order of activity among alkaline earth oxide catalysts is BaO > SrO > CaO > MgO. CaO is the most frequently applied metal oxide catalyst for biodiesel preparation, due to its cheap price, relatively high basic strength and less environmental impacts. Reddy et al. (2006) used CaO as solid base catalyst in the transesterification of soybean oil, only 2% biodiesel yield achieved. However, the intrinsic basicity of nano-CaO is much higher, 99% biodiesel yield was obtained with nano-CaO. In addition to specific surface area, other variables such as temperature and molar ratio of methanol to oil also influenced the catalytic activity. A measured amount of water in oil is wonderful for promotion of catalytic activity. Study showed 95% biodiesel yield was obtained using CaO as catalyst in present of about 2 wt% water (Xi et al., 2008). SrO has basic sites stronger than $H_0 = 26$. It can catalyze many chemical reactions, such as oxidative coupling of methane, selective oxidation of propane, nitro-Aldol reactions and mixed Tishchenko reactions. Liu et al (2007) reported SrO performed high catalytic activity to convert soybean oil to biodiesel with a yield over 95% at temperature below 70 ºC for 30 min. MgO has weak basic strength and low solubility in methanol. It is usually produced by direct heating of magnesium carbonate or magnesium hydroxide. There is a striking linear correspondence between the catalytic activity and surface basicity of MgO. High reaction temperature (e.g., 523 ºC) and high pressure (e.g., 24 MPa) are usually needed for achieving a high biodiesel yield (Wang and Yang, 2007).

To increase the basic strength of a single metal oxide, mixed metal oxides are synthesized. Peterson et al. (1984) prepared CaO-MgO and found that it provided higher catalytic activity than CaO powders for transforming rapeseed oil to biodiesel. Catalytic activity tests were performed for $CaO-TiO_2$, $CaO-MnO_2$, $(CaO)_2-Fe_2O_3$, $CaO-ZrO_2$ and $CaO-CeO_2$ samples, approximately 90% biodiesel yield were obtained (Kawashima et al., 2008). The Ca catalysts were found to have higher basicity and activity. Such catalysts performed noticeably decreased activity in transesterificaiton when ethanol or branched alcohols was used, attributed to the steric effects on the catalytic activity of these catalysts. Furthermore, active sites of metal oxides are easily blocked by adsorbing intermediates (diglyceride, monoglyceride) or products. Deactivated catalysts can be recovered nearly to the initial value through calcination.

2.2.3 Metallic salts

Inorganic solid bases, such as sodium silicate (Guo et al., 2010), vanadyl phosphate (Serio et al., 2007), calcium zincate (Rubio-Caballero et al., 2009) and calcium methoxide (Liu et al., 2008), are low-cost and easy-to-use heterogeneous catalysts. Reports on metallic salts catalyzed conversion in biodiesel preparation are rare. Here, only sodium silicate, vanadyl phosphate (VOP) and calcium zincate are reviewed.

Sodium silicate was used as starting materials to synthesize γ-zeolite, NaY zeolite, and NaX zeolite. Guo et al. (2010) used sodium silicate to catalyze the transesterification reaction for the first time. It catalyzed soybean oil to biodiesel with a yield of almost 100% under the conditions: sodium silicate of 3.0 wt %, a molar ratio of methanol/oil of 7.5:1, reaction time of 60 min, reaction temperature of 60 ºC, and stirring rate of 250 rpm. In addition to high catalyst activity, sodium silicate also has other similar characteristics to supported-solid base catalysts. Most of basic sites were in the interior of the solid catalyst due to low surface area and high density of the basic sites. The calcined sodium silicate could tolerate 4.0 wt% water or 2.5 wt% FFAs contained in soybean oil. The water tolerance is related to its special crystal and porous structure. In the presence of high amount of water, a sequential hydration will occur in three steps:

$$\equiv Si - O - Na + H_2O \rightarrow \ \equiv Si - O - H + OH^-$$
$$\equiv Si - O - Si \equiv +OH^- \rightarrow \ \equiv Si - O - H + \equiv Si - O^- \qquad (1)$$
$$\equiv Si - O^- + H_2O \rightarrow \ \equiv Si - O - H + OH^-$$

As a result, Si-O-Si bridges would be hydrolyzed and H_4SiO_4 monomers are sequentially released. Such series of reactions not only produce OH-, but also avoid the formation of soap. Furthermore, sodium silicate could also be used to catalyze dehydration of glycerol. Long et al. (2011) used sodium silicate as catalyst for transesterification of rapeseed oil for several recycles, and subsequently the used sodium silicate without any modification was catalyzed for the hydrothermal production of lactic acid from glycerol at 300 °C. A yield of 80.5% lactic acid and only minor amounts of formic, acetic acid and acrylic acid were produced.

Previous applications of VOP were mainly in hydrocarbon oxidation, dehydration and isomerization (Serio et al., 2006). Serio et al. (2007) confirmed that VOP-based catalysts were very active in the transesterification of vegetable oil with methanol despite their low specific surface area. VOP was deactivated due to a progressive reduction of vanadium (V) species from V^{5+} to V^{4+} and V^{3+} by methanol. Because the deactivation is reversible and catalyst activity can easily be restored by calcination. Rubio-Caballero et al. (2009) investigated the use of calcium zincate in the methanolysis of sunflower oil for biodiesel production. The activated calcium zincate at 400 °C is stable against lixiviation, attributed to its strong interaction with a much less soluble zinc oxide. But, calcium zincate is more sensitive to water (> 0.2 wt.%) rather than FFAs. Calcium methoxide has a moderate surface area, relative broad particle size distribution, narrow pore size distribution, strong basicity, long catalyst lifetime and better stability in organic solvent (Liu et al., 2008). It has tremendous potential to replace some homogeneous catalysts.

2.2.4 Supported base catalysts

Alkali metals (Li, Na, K) and alkaline earth metals (Mg, Ca, Ba) are the most common sources of super basicity, and selected as the active species of supported catalysts for biodiesel synthesis. They are frequently used in the metallic form or as various ionic forms of hydroxide, halide, carbonate and nitrate, such as K^+, Li^+ La^{3+}, KOH, NaOH, KF, K_2CO_3, KNO_3 (Shu et al., 2007; Sun et al., 2008; Vyas et al., 2009). Alumina, silica, zinc oxide, zirconium oxide and zeolite were used as supports for these catalysts. Surface basicity is the primary determinant of catalyst activity, then the specific surface area and pore volume (Sun et al., 2008). During the preparation of such catalysts, the mechanical intensity and surface area of carriers can be adjusted to obtain different basic intensities and activity sites.

Almost all supported base catalysts were synthesized via loading of active species on carriers by covalent bond, ionic bond or physical adsorption. Despite of formation of M-O-carrier (e.g., Al-O-K, Si-O-Na and Ca-O-K), other possible interactions of the alkali species with supports include formation of solid solutions and acid-base reactions. Hydroxyl groups introduced to the surface of solids play an important role in transesterification reaction (Xie et al. 2006). The hydroxyls with alkali species enhance the catalytic activity.

As a most popular carrier, Al_2O_3 has almost all noteworthy properties such as high temperature resistant, high surface area, high porosity, low density and transition crystalline phase existed in a wide temperature range. Furthermore, it serves as carrier with both solid acid and base. Most super basicity sources can be well dispersed on the Al_2O_3 support in the form of a monolayer at a low loading. Furthermore, alumina is more resistant than other supports (e.g., SiO_2, CaO and zeolite) for alkali species. Taking KNO_3/Al_2O_3 as an example, it is usually prepared by impregnation and subsequent calcination at 500 °C (Xie et al. 2006). K^+ ions replaced protons of isolated hydroxyl groups to form Al-O-K groups. The Al-O-K groups and K_2O derived from KNO_3 are active basic species. The base strength could be tentatively denoted as $15 < H_0 < 18.4$ by using Hammett indicator. Basic strength of KNO_3/Al_2O_3 was influenced by KNO_3 loading and temperature. The 35% KNO_3/Al_2O_3 sample calcined at 500 °C had the highest basicity. However, the sample prepared at 700 °C was most stable. Because part of potassium species are loss by a solid-solid reaction leading to formation of spinels or penetration into the subsurface.

Aends and Sheldon (2001) indicated that such kind of catalyst is unstable during reaction, mainly due to M-O-Al decomposed in present of methanol. Arzamendi et al (2007) confirmed that NaOH reacted with the support to form aluminates during preparation of $NaOH/Al_2O_3$. Leaching of sodium species from Al_2O_3 was also found. Furthermore, problems of high cost, difficult preparation and easy poisoning by absorption of H_2O and CO_2 should be solved. The supported solid base catalysts are excellent for transesterification of triglyceride, but a higher temperature is needed.

3. Catalytic mechanism

3.1 Heterogeneous solid acid-catalyzed esterification mechanism
Low-cost feedstocks need pretreatment (esterification) to remove FFAs before base-catalyzed transesterification reaction. The esterification path is relatively simple reversible reaction as follows:

$$\text{(2)}$$

In the reaction (2), FFA is converted to FAME. When homogenous acid (e.g., sulfonate acid, phosphorus acid and hydrochloric acid) was used, esterification reaction is a process that FFA supply hydroxide and methanol supply proton without intermediate process.

Different to homogeneous catalysis, heterogeneous catalytic process is known to follow a carbonium ion mechanism. The mechanism of solid acid-catalyzed esterification consists of following steps as shown in Fig. 3. Firstly, solid catalysts provided protons, and carbonyl carbon was protonated. Next, nucleophilic attack of CH_3OH on the carbonium ion formed a tetrahedral intermediate. Finally, FAME was produced after proton migrated and the intermediate broke down, and proton was reformed.

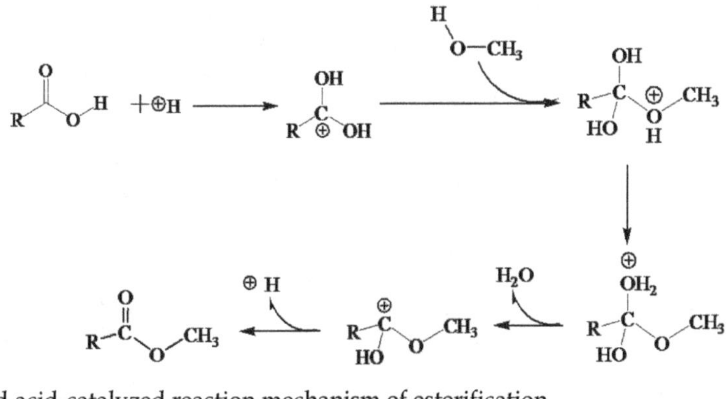

Fig. 3. Solid acid-catalyzed reaction mechanism of esterification.

The esterification reaction path is slightly different in various acidic species types. The whole reaction process is through proton-exchange. Tesser et al. (2005) proposed a kinetic model based on the following hypotheses: (1) major part of the active sites are occupied by methanol in a protonated form, and the rest part are also occupied; (2) fatty acid, water and methyl ester reach proton-exchange equilibrium with the protonated methanol; (3) inside the resin particles, an Eley-Rideal mechanism occurs between protonated fatty acid and the methanol. Deviate from the mechanism shown in Fig. 3, steps of protonation of carbonyl carbon, nucleophilic attack, proton migration and breakdown of intermediate are undergoing in a proton-exchange way.

3.2 Transesterification mechanism

The transesterification reaction involves catalytic reaction between triglyceride and alcohol (e.g., methanol, ethanol, propanol and butanol) to form biodiesel (FAMEs) and glycerol (Fig. 4). In the reaction, three consecutive reactions are required to complete the transesterification of a triglyceride molecule. In the presence of acid or base, a triglyceride molecule reacts with an alcohol molecule to produce a diglyceride and FAME. Then, a diglyceride reacts with alcohol to form a monoglyceride and FAME. Finally, an monoglyceride reacts with alcohol to form FAME and glycerol. Diglyceride and monoglyceride are the intermediates in this process.

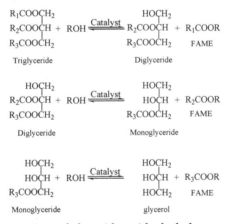

Fig. 4. Transesterification reactions of glycosides with alcohol.

3.2.1 Mechanism for heterogeneous acid-catalyzed transesterification

Acidic or basic functional groups in the active sites of solid catalysts catalyze the reaction by donating or accepting protons. Acid-catalyzed reaction mechanism for the transesterification of triglycerides is shown in Fig. 5. Firstly, triglycerides are protonated at the carbonyl group on the surface of solid acid. Then, a nucleophilic attack of the alcohol to carbocation forms a tetrahedral intermediate (hemiacetal species). Unstable tetrahedral intermediate leads to proton migration, followed by breakdown of the tetrahedral intermediate with assistance of solvent. After repeating twice, three new FAME as products were produced and the catalyst was regenerate. During the catalytic process, protonation of carbonyl group boosts the catalytic effect of solid acid catalyst by increasing the electrophilicity of the adjacent carbonyl carbon atom.

Different with Brønsted acids, Lewis acids [e.g., $Fe_2(SO_4)_3$, titanate complexes, carboxylic salts, divalent metal pyrone] act as electron-acceptors via the formation of a four-membered ring transition state (Abreu et al., 2004; Di Serio et al., 2005). The reactant triglyceride and metal form a Lewis complex, which assists solid Lewis acids during process of the carbonyl groups activating for a nucleophilic attack by the reactant alcohol. The triglyceride carbonyl coordinates at a vacant site in the catalytic active specie. Formation of a more electrophilic species is responsible for the catalytic activity. Stearate metals (Ca, Ba, Mg, Cd, Mn, Pb, Zn, Co and Ni) were tested as catalysts for methanolysis of soybean oil (2.0 g) with methanol (0.88 g) at 200 °C (Di Serio et al., 2005). A high FAMEs yield (96%) and a low final FFAs concentration (<1%) were obtained in a relatively short reaction time (200 min).

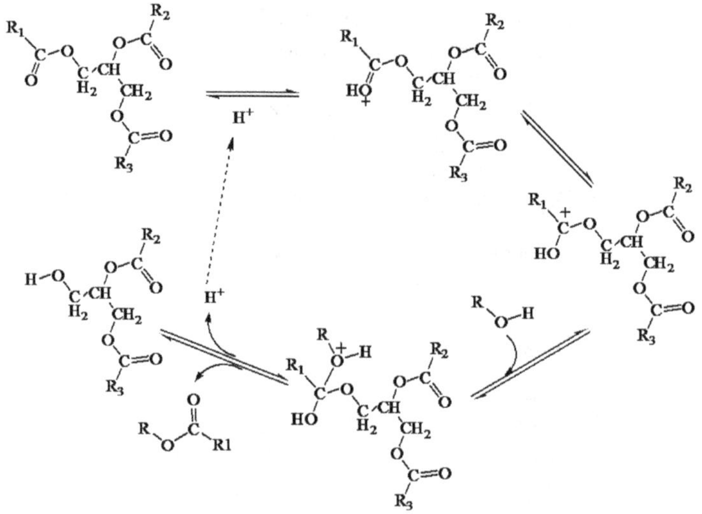

Fig. 5. Acid-catalyzed reaction mechanism of transesterification.

3.2.2 Mechanism for heterogeneous base-catalyzed transesterification

Base-catalyzed crude oil to biodiesel gets more studies than acid-catalyzed method. In base-catalyzed process, OH- or CH_3O- ions performed as active species. Catalytic reactions started on the surface of heterogeneous base (Fig. 6). The mechanistic pathway for solid base-catalyzed transesterification seems to follow a similar mechanism to that of a homogeneous base catalyst. First, ion-exchange proceeded after methanol absorbed on the surface of solid base, producing catalytic active specie (CH_3O-) which is strongly basic and highly catalytic active. Secondly, nucleophilic attack of CH_3O- on the carbonyl carbon of triglyceride formed a tetrahedral intermediate. Thirdly, rearrangement of the intermediate resulted in the formation of FAME. Finally, protons were converted to diglyceride ion to generate diglyceride. This sequence was then repeated twice to yield glycerol and biodiesel.

Formation of CH_3O- is different according to solid base types. Taking CaO as an example, surface O^{2-} is the basic site, which can extract H^+ from H_2O to form OH-, and OH- extracts H^+ from methanol to generate CH_3O- (Liu et al., 2008). It is interesting that CaO generates more methoxide anions in the presence of a little water (less than 2.8% by weight of crude oil), avoiding formation of soap. Surface oxides or hydroxide groups depend on the basicity and catalytic activities. The basic strengths of Na/CaO and K/CaO are slightly lower than that of Li/CaO (Ma and Hanna, 1999). The presence of the electron-deficient M^+ on the support enhances the basicity and activity of the catalysts towards the transesterification reaction.

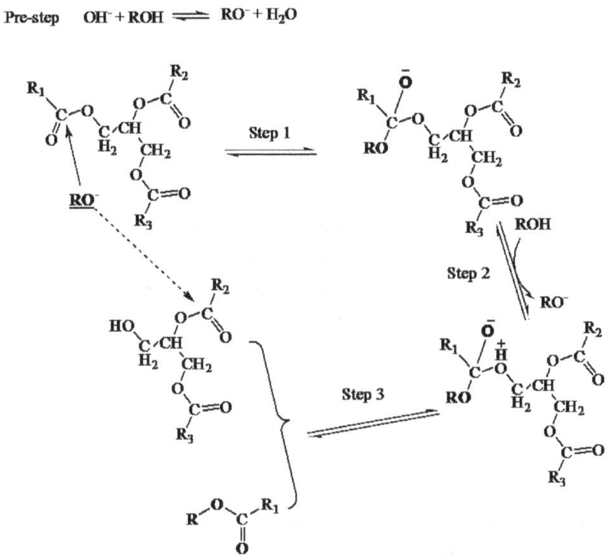

Fig. 6. Base-catalyzed reaction mechanism of transesterification.

4. Other methods or technologies

4.1 Microwave technology

Microwave heating has been widely used in many areas to affect chemical reaction pathways and accelerate chemical reaction rates. Microwave irradiation can accelerate the chemical reaction, and high product yield can be achieved in a short time. Microwave irradiation assisted biodiesel synthesis is a physicochemical process since both thermal and non-thermal effects are often involved, which activates the smallest degree of variance of polar molecules and ions such as alcohol with the continuously changing magnetic field. Upon microwave heating, rapid rising of temperature would result in interactions of changing electrical field with the molecular dipoles and charged ion, leading to a rapid generation of rotation and heat due to molecular friction. Dielectric properties are important in both the design calculations for high frequency and microwave heating equipment. Furthermore, dielectric constant depends on frequency, and is strongly influenced by temperature, mixed ratio and solvent type.

In Azcan and Danisman's work (2007), microwave heating effectively reduced reaction time from 30 min (for a conventional heating system) to 7 min. Ozturk et al. (2010) studied microwave assisted transesterification of maize oil, using a molar ratio alcohol/maize-oil of 10:1, and 1.5% w/w NaOH as catalyst. A 98.3% conversion rate is obtained using methanol for 5 min. Based on special heating manner, microwave irradiation performed well in transesterification of vegetable oil with heterogeneous base. Hsiao et al. (2011) introduced nano-powder calcium oxide as solid base in converting soybean oil to biodiesel. A 96.6% of conversion rate was obtained under conditions of methanol/oil molar ratio of 7:1, amount of catalyst of 3.0 wt.%, reaction temperature of 65 °C and reaction time of 60 min. While a biodiesel conversion rate exceeded 95% was achieved under conditions of 12:1 molar ratio of methanol to oil, 8 wt.% catalyst, 65 °C reaction temperature and 2.0% water content for 3 h (Xie et al., 2008). Microwave irradiation is also used for extraction of bioactive compounds for value-added products, including oil extraction systems. Microwave heating can be used for biodiesel production by in-situ simultaneous extraction and transesterification from oil seeds.

4.2 Ultrasonic technology

There are three primary effects on an object under ultrasound: (1) Mechanical effects; (2) Cavity effects; (3) Thermal effects. The above effects of ultrasound not only change the structure of the object, but also lead to chemical reactions. Ultrasonic radiation is a relative new technique that results in the formation and collapse of micro-scale bubbles in liquid to generate local high temperature and high pressure. So, it is used as alternative energy source to promote reactions. The cavitation in ultrasonic wavelength is the phenomenon of expansion and contraction of the transfer media bubbles. Ultrasonic energy is propagated into solution by the destruction of pressurized micro-bubbles into small droplets. Furthermore, ultrasonication device placed near the liquid–liquid interface in a two-phase reaction system benefited for producing large interfacial areas (Wu et al., 2007). Cavitation induced by ultrasound has significant effects on liquid phase reactions. When ultrasound irradiation increased from 30 to 70 W, the mean droplet size decreased from 156 nm to 146 nm. Nevertheless, effect of droplet size on biodiesel yield was not studied.

Ultrasound has a short wavelength, slow transfer rate, and high energy transmittance as the vibrating type energy. Irradiation of ultrasonic energy has been used for the (trans)esterification of vegetable oils to shorten reaction time and to increase product yield (Deng et al., 2010). A comparison study between conventional and ultrasonic preparation of beef tallow biodiesel was carried out (Teixeira et al., 2009). The results showed that conversion rate and biodiesel quality were similar. The use of ultrasonic irradiation decreased reaction time from 1 h to 70 s. In addition to the mentioned advantages, ultrasonic can promote the deposition of glycerol at the bottom of reactor. Stavarache et al. (2007) investigated a bench-scale continuous process for biodiesel synthesis from neat vegetable oils under high power, low frequency ultrasonic irradiation. Reaction time and alcohol-oil molar ratio were mainly variables affecting the transesterification. Their research confirmed that ultrasonic irradiation is suitable for large-scale processing of vegetable oils since relatively simple devices can be used to perform the reaction. In the process, however, real irradiation time decreased during increasing pulse interval for tuning temperature, leading to biodiesel yield decrease. To reduce the effect of irradiation time loss, reaction temperature should be kept constant.

Mass transfer resistance is one of the main reasons for poor catalytic performance of solid catalysts in (trans)esterification. Very fine ultrasonic emulsions greatly improve the interfacial area available for reaction, increase the effective local concentration of reactive species, and enhance the mass-transfer in interfacial region. Therefore it leads to a remarkable increase in reaction rate under phase-transfer conditions transesterification with solid catalyst. Ultrasonication could reduce the transesterification reaction time to around 10 min compared with over 6 h for conventional processing.

4.3 Ionic liquids

Ionic liquids (ILs) are defined as salts that are in the state of liquid at low temperatures (below 100 °C). They are composed solely of cations and anions, and were used as solvents/catalysts for reactions. ILs are nonvolatile and thermal stable, hence they are excellent alternatives to traditional solvents. Some ILs are Lewis and Franklin acids. Acidic ILs are new-type of catalysts with high-density active sites as liquid acids but non-volatilization as solid acids. Furthermore, cations and anions of ILs can be designed to bind a series of groups with specific properties, so as to achieve the purpose of regulating the acidity. Recently, they have been used to replace traditional liquid acids such as sulfuric acid and hydrochloric acid for biomass conversion (Qi et al., 2010).

ILs were originally used as solvents for biodiesel synthesis with high biodiesel yield in short reaction time, by forming an effective biphasic catalytic system for the transesterification reaction. Neto et al. (2007) introduced a complex [Sn(3-hydroxy-2-methyl-4-pyrone)$_2$(H$_2$O)$_2$] immobilized in BMI·InCl$_4$ with high price metal salts, and a maximum biodiesel yield of 83% was achieved. Later, biodiesel synthesis from vegetable oils using imidazolium-based ionic liquids under multiphase acidic and basic conditions was reported (Lapis et al., 2008). It is found that the acid is almost completely retained in ionic liquid phase, and ILs could be reused at least six times without any significant loss in the biodiesel yield or selectivity. However, the ILs is expensive and was only used for neutral vegetable oils. Brønsted acidic ILs were highly efficient catalysts for biodiesel synthesis from vegetable oils. Sulfuric acid groups in these ILs are the active sites for transesterification. Dicationic ILs exhibited better stability than the traditional ones. The acidic dicationic ILs with an alkane sulfuric acid group gave a superior catalytic performance in esterification reaction. Neto et al. (2007) assumed that the use of ILs with inherent Lewis acidity may constitute a more stable and robust catalytic system for the transesterification reaction. Guo et al. (2011) used 7 low-cost commercial ILs as both catalysts and solvents for the direct production of biodiesel from un-pretreated *Jatropha* oil. It was found that [BMIm][CH$_3$SO$_3$] had the highest catalytic activity with 93% of oleic acid being converted into ethyl oleate. When FeCl$_3$ was added to [BMIm][CH$_3$SO$_3$], a maximum biodiesel yield of 99.7% was achieved from un-pretreated *Jatropha* oil. However, it is complicated to synthesize these functional ILs and their cost is too high for industrial applications. Therefore, further investigation is necessary to synthesize inexpensive, stable and highly-active ILs.

5. Conclusions and future perspectives

Currently, homogeneous catalysis is a predominant method for transesterification reaction. Separating the catalyst from a mixture of reactants and product is technically difficult. Compared with liquid acid catalysts, solid acid catalysts have distinct advantages in recycling, separation, and environmental friendliness. Solid acid catalysts are easily separated from the products mixture for reuse after reaction. Both Lewis acid–base sites and Brønsted acid-base sites have the ability to catalyze oil transesterification reaction. Besides specific surface area, pore size and pore volume, the active site concentration and acidic type are important factors for solid acid performance. Moreover, types of active precursor have significant effect on the catalyst activity of supported catalysts. However active site concentration was found to be the most important factor for solid catalyst performance. Solid acids with a large potential for synthesis of biodiesel should have a large number of Brønsted acid sites and good thermal stability. A good solid catalyst with sufficient catalytic

activity combined with appropriate reactor design should make it possible to realize biodiesel production on a practical scale.
Among solid catalysts introduced in this chapter, Solid acid (i.e. ion-exchange resins, HPAs and supported acid catalysts) and Solid base (i.e. hydrotalcites, metallic salts and supported base catalysts) are promising material for study. Low-cost catalysts that still retain the advantages of a supported base catalyst should be developed to simplify the preparation process. Design of solid catalysts with higher activity is an important step for clean production of biodiesel. Innovation and breakthrough in hydrolysis process is a key for commercialization of solid acid catalysts. In the near future, through the combination of green solvents, chemical process, biotechnology and catalysis, it can be expected that novel solid catalysts will replace the current-used homogeneous catalysts in biodiesel peoduction.

6. References

Abreu, F.R., Lima, D.G., Hamúa, E.H., Wolf, C., Suarez, P.A.Z. (2004) Utilization of metal complexes as catalysts in the transesterification of Brazilian vegetable oils with different alcohols. *Journal of Molecular Catalysis A: Chemical*, Vol.209, No.1-2, pp. 29-33.

Aends, I.W.C.E., Sheldon, R.A. (2001) Activities and stabilities of heterogeneous catalysts in selective liquid phase oxidations: recent developments. *Applied catalysis A: General*, Vol.212, No.1-2, pp. 175-187.

Alsalme, A., Kozhevnikova, E.F., Kozhevnikov, I.V. (2008) Heteropoly acids as catalysts for liquid-phase esterification and transesterification. *Applied Catalysis A: General*, Vol.349, pp. 170-176.

Arzamendi, G., Campo, I., Arguínarena, E., Sánchez, M., Montes, M., Gandá, L.M. (2007) Synthesis of biodiesel with heterogeneous NaOH/alumina catalysts: Comparison with homogeneous NaOH. *Chemical Engineering Journal*, Vol.134, pp 123-130.

Azcan, N., Danisman, A. (2007) Alkali catalyzed transesterification of cottonseed oil by microwave irradiation. *Fuel*, Vol.86, pp. 2639-2644.

Bancquart, S., Vanhove, C., Pouilloux, Y., Barrault, J. (2001) Glycerol transesterification with methyl stearate over solid basic catalysts: I. Relationship between activity and basicity. *Applied Catalysis A: General*, Vol.218, pp 1-11.

Brito, A., Borges, M.E., Garín, M., Hernández, A. (2009) Biodiesel production from waste oil using Mg-Al layered doublehydrotalcite catalysts. Energy&Fuels, Vol.23, pp. 2952-2958.

Caetano, C.S., Fonseca, I.M., Ramos, A.M., Vital, J., Castanheiro, J.E. (2008) Esterification of free fatty acids with methanol using heteropolyacids immobilized on silica. *Catalysis Communications*, Vol.9, pp. 1996-1999.

Caetano, C.S., Guerreiro, L., Fonseca, I.M., Ramos, A.M., Vital, J., Castanheiro, J.E. (2009) Esterification of fatty acids to biodiesel over polymers with sulfonic acid groups. *Applied Catalysis A: General*, Vol.359, No.1-2, pp. 41-46.

Cantrell, D.G., Gilie, L.J., Lee, A.F., Wilson, K. (2005) Structure-reactivity correlations in Mg-Al hydrotalcite catalysts for biodiesel synthesis. *Applied Catalysis A: General*, Vol.287, No. 2, pp. 183-190.

Chung, K.H., Chang, D.R., Park, B.G. (2008) Removal of free fatty acid in waste frying oil by esterification with methanol on zeolite catalysts. *Bioresource Technology*, Vol.99, pp. 7438-7443.

Deng, X., Fang, Z., Liu, Y.H. (2010) Ultrasonic transesterification of Jatropha curcas L. oil to biodiesel by a two-step process. *Energy Conversion and Management*, Vol.51, pp. 2802-2807.

Deng, X., Fang, Z., Liu, Y.H., Yu, C.L. (2011) Production of biodiesel from Jatropha oil catalyzed by nanosized solid basic catalyst. *Energy*, Vol.36, No.2, pp. 777-784.

Di Serio, M., Dimiccoli, M., Cammarota, F., Nastasi, M., Santacesaria, E. (2005) Synthesis of biodiesel via homogeneous Lewis acid catalyst. *Journal of Molecular Catalysis A: Chemical*, Vol.239, pp.111-115.

Feng, Y.H., He, B.Q., Cao, Y.H., Li, J.X., Liu, M., Yan, F., Liang, X.P. (2010) Biodiesel production using cation-exchange resin as heterogeneous catalyst. *Bioresource Technology*, Vol.101, pp. 1518-1521.

Feng, Y.H., Zhang, A.Q., Li, J.X., He, B.Q. (2011) A continuous process for biodiesel production in a fixed bed reactor packed with cation-exchange resin as heterogeneous catalyst. *Bioresource Technology*, Vol. 102, pp.3607-3609.

Garcia, C.M., Teixeira, S., Marciniuk, L.L., Schuchardt, U. (2008) Transesterification of soybean oil catalyzed by sulfated zirconia. *Bioresource Technology*, Vol. 99, pp. 6608-6613.

Giri, B.Y., Rao, K.N., Devi, B.L.A.P., Lingaiah, N., Suryanarayana, I., Prasad, R.B.N., Prasad, P.S.S. (2005) Esterification of palmitic acid on the ammonium salt of 12-tungstophosphoric acid: The influence of partial proton exchange on the activity of the catalyst. *Catalysis Communications*, Vol. 6, pp. 788-792.

Guo, F., Fang, Z., Tian, X.F., Long, Y.D., Jiang, L.Q. (2011) One-step production of biodiesel from Jatropha oil with high-acid value in ionic liquids. *Bioresource Technology*, Vol.102, No.11, pp.6469-6472.

Guo, F., Peng, Z.G., Dai, J.Y., Xiu, Z.L. (2010) Calcined sodium silicate as solid base catalyst for biodiesel production. *Fuel Processing Technology*, Vol. 91,No.33, pp.322-328.

Hsiao, M.C., Lin, C.C., Chang, Y.H. (2011) Microwave irradiation-assisted transesterification of soybean oil to biodiesel catalyzed by nanopowder calcium oxide. *Fuel*, Vol.90, pp.1963-1967.

Karmee, K.S., Chadha, A. (2005) Preparation of biodiesel from crude oil of *Pongamia pinnata*. *Bioresource Technology*, Vol.96, pp.1425-1429.

Kawashima, A., Matsubara, K., Honda, K. (2008) Development of heterogeneous base catalysts for biodiesel production. *Bioresource Technology*, Vol.99, pp.3439-3443.

Kozhevnikov, I.V. (1998) Catalysis by heteropoly acids and multicomponent polyoxometalates in liquid-phase reactions. *Chemical Reviews*, 98: 171-198.

Lapis, A.A., Oliveira, L.F., Neto, B.A.D., Dupont, J. (2008) Ionic liquid supported acid/base-catalyzed production of biodiesel. *ChemSusChem*, Vol.1, pp.759-762.

Lee, D.W., Park, Y.M., Lee, K.Y. (2009) Heterogeneous Base Catalysts for Transesterification in biodiesel synthesis. *Catal Surv Asia*, Vol.13, pp.63-77.

Liu, X., He, H., Wang, Y., Zhu, S.L. (2007) Transesterification of soybean oil to biodiesel using SrO as a solid base catalyst. *Catalysis Communications*, Vol.8, pp. 1107-1111.

Liu, X.J., He, H.Y., Wang, Y.J., Zhu, S.L., Piao, X. (2008) Transesterification of soybean oil to biodiesel using CaO as a solid base catalyst. *Fuel*, Vol.87, pp.216-221.

Liu, X.J., Piao, X.G., Wang, Y.J., Zhu, S.L., He, H.Y. (2008) Calcium methoxide as a solid base catalyst for the transesterification of soybean oil to biodiesel with methanol. *Fuel*, Vol.87, pp.1076-1082.

Liu, Y., Wang, L., Yan, Y. (2009) Biodiesel synthesis combining pre-esterification with alkali catalyzed process from rapeseed oil deodorizer distillate. *Fuel Processing Technology*, Vol.90, pp. 857-862.

Lotero, E., Liu, Y.J., Lopez, D.E. (2005) Suwannakarn K, Bruce DA, Goodwin JG. Synthesis of biodiesel via acid catalysis. *Industrial & Engineering Chemistry Research*, Vol.44, No.14, pp. 5353-5363.

Long, Y.D., Guo, F., Fang, Z., Tian, X.F., Jiang, L.Q., Deng, X., Zhang, F. (2011) Production of biodiesel and lactic acid from rapeseed oil using sodium silicate as catalyst. *Bioresource Technology*, Vol.102, pp.6884-6886.

Ma, F., Hanna, M.A. (1999) Biodiesel production: a review. *Bioresource Technology*, 70: 1-15.

Montero, J.M., Gai, P., Wilson, K., Lee, A.F. (2009) Structure-sensitive biodiesel synthesis over MgO nanocrystals. *Green Chemistry*, Vol.11, pp. 265-268.

Narasimharao, K., Brown, D.R., Lee, A.F., Newman, A.D., Siril, P.F., Tavener, S.J., Wilsona, K. (2007) Structure-activity relations in Cs-doped heteropolyacid catalysts for biodiesel production. *Journal of Catalysis*, Vol.248, pp.226-234.

Neto, B.A.D., Alves, M.B., Lapis, A.A.M., Nachtigall, F.M., Eberlin, M.N., Dupont, J., Suarez, P.A.Z. (2007) 1-n-Butyl-3-methylimidazolium tetrachloro-indate (BMI·InCl₄) as a media for the synthesis of biodiesel from vegetable oils. *Journal of Catalysis*, Vol.249, pp.154-161.

Noiroj, K., Intarapong, P., Luengnaruemitchai, A., Jai-In, S. (2009) A comparative study of KOH/Al₂O₃ and KOH/NaY catalysts for biodiesel production via transesterification from palm oil. *Renewable Energy*, Vol.34, pp.1145-1150.

Okuhara, T., Nishimura, T., Watanabea, H., Misono, M. (1992) Insoluble heteropoly compounds as highly active catalysts for liquid-phase reactions. *Journal of Molecular Catalysis*, Vol.74, No.1-3, pp. 247-256.

Ozturk, G., Kafadar, A.B., Duz, M.Z., Saydut, A., Hamamci, C. (2010) Microwave assisted transesterification of maize (*Zea mays L.*) oil as a biodiesel fuel. *Energy Exploration & Exploitation*, Vol.28, No.1, pp.47-57.

Peterson, G.R., Scarrach, W.P. (1984) Rapeseed oil transester- ification by heterogeneous catalysis. *Journal of the American Oil Chemists' Society*, Vol.61, pp.1593-1596.

Pérez-Pariente, J., Díaz, I., Mohino, F., Sastre, E. (2003) Selective synthesis of fatty monoglycerides by using functionalised mesoporous catalysts. *Applied Catalysis A: General*, Vol.254, pp.173-188.

Qi, X.H., Watanabe, M., Aida, TM., Smith Jr., R.L. (2010) Efficient one-pot production of 5-hydroxymethylfurfural from inulin in ionic liquids. *Green Chemistry*, Vol.12, pp.1855-1860.

Reddy, C.R.V., Oshel, R., Verkase, J.G. (2006) Room-temperature conversion of soybean oil and poultry fat to biodiesel catalyzed by nanocrystalline calcium oxides. *Energy & Fuels*, Vol.20, No.3, pp.1310-1314.

Rubio-Caballero, J.M., Santamaría-González, J., Moreno-Tost, R., Jiménez-López, A., Maireles-Torres, P. (2009) Calcium zincate as precursor of active catalysts for biodiesel production under mild conditions. *Applied Catalysis B: Environmental*, Vol.91, pp.339-346.

Russbueldt, B.M.E., Hoelderich, W.F. (2009) New sulfonic acid ion-exchange resins for the preesterification of different oils and fats with high content of free fatty acids. *Applied Catalysis A: General*, Vol.362, pp.47-57.

Sasidharan, M., Kumar, R. (2004) Transesterification over various zeolites under liquid-phase conditions. *Journal of Molecular Catalysis A: Chemical*, Vol.210, pp.93-98.

Serio, M.D., Cozzolino, M., Tesser, R., Patrono, P., Pinzarij Bonelli, B., Santacesaria, E. (2007) Vanadyl phosphate catalysts in biodiesel production. *Applied Catalysis A: General*, Vol.320, pp.1-7.

Serio, M.D., Ledda, M., Cozzolino, M., Minutillo, G., Tesser, R., Santacesaria, E. (2006)Transesterification of soybean oil to biodiesel by using heterogeneous basic catalysts. *Industrial & Engineering Chemistry Research*, Vol.45, No.9, pp.3009-3014.

Shibasaki-Kitakawa, N., Honda, H., Kuribayashi, H., Toda, T., Fukumura, T., Yonemoto, T. (2007) Biodiesel production using anionic ion-exchange resin as heterogeneous catalyst. *Bioresource Technology*, Vol.98, pp. 416-421.

Shokrolahi, A., Zali, A., Pouretedal, H.R., Mahdavi, M. (2008) Carbon-based solid acid catalyzed highly efficient oxidations of organic compounds with hydrogen peroxide. *Catalysis Communications*, Vol.9, pp. 859–863.

Shu, Q., Yang, B., Yuan, H., Qing, S., Zhu, G. (2007) Synthesis of biodiesel from soybean oil and methanol catalyzed by zeolite beta modified with La^{3+}. *Catalysis Communications*, Vol.8, pp.2159-2165.

Shu, Q., Zhang, Q., Xu, G.H., Nawaz, Z., Wang, D.Z., Wang, J.F. (2009) Synthesis of biodiesel from a model waste oil feedstock using a carbon-based solid acid catalyst: reaction and separation. *Fuel Processing Technology*, Vol.90, pp.1002-1008.

Son, S.M., Kimura, H., Kusakabe, K. (2011) Esterification of oleic acid in a three-phase, fixed-bed reactor packed with a cation exchange resin catalyst. *Bioresource Technology*, Vol.102, pp.2130-2132.

Stavarache, C., Vinatoru, M., Bandow, H. (2007) Ultrasonically driven continuous process for vegetable oil transesterification. *Ultrasonics Sonochemistry*, Vol.14, pp.413-417.

Suganuma, S., Nakajima, K., Kitano, M., Yamaguchi, D., Kato, H., Hayashi, S., Hara, M. (2008) Hydrolysis of cellulose by amorphous carbon bearing SO_3H, COOH, and OH groups. *JACS*, Vol.130, pp.12787-12793.

Sun, H., Hu, K., Lou, H., Zheng, X.M. (2008) Biodiesel production from transesterification of rapeseed oil using KF/Eu_2O_3 as a catalyst. *Energy Fuels*, Vol.22, No.4, pp.2756-2760.

Teixeira, L.S.D., Assis, J.C.R., Mendonça, D.R., Santos, I.T.V., Guimarães, P.R.B., Pontes, L.A.M., Teixeira, J.S.R. (2009) Comparison between conventional and ultrasonic preparation of beef tallow biodiesel. *Fuel Processing Technology*, Vol.90, pp.1164-1166.

Tesser, R., Serio, M.D., Carotenuto, G., Santacesaria, E. (2010) Absorption of water/methanol binary system on ion-exchange resins. *Canadian journal of chemical engineering*, Vol.88, No.6, pp.1044-1053.

Toda, M., Takagaki, A., Okamura, M., Kondo, J.N., Hayashi, S., Domen, K., Hara, M. (2005) Biodiesel made with sugar catalyst. *Nature*, Vol.438, pp.178.

Vyas, A.P., Subrahmanyam, N., Patel, P.A. (2009) Production of biodiesel through transesterification of Jatropha oil using KNO_3/Al_2O_3 solid catalyst. *Fuel*, Vol.88, No.4, pp. 625-628.

Wang, L., Yang, J. (2007) Transesterification of soybean oil with nano-MgO or not in supercritical and subcritical methanol. *Fuel*, Vol.86, pp. 328-333.

Wu, P., Yang, Y., Colucci, J.A., Grulke, E.A. (2007) Effect of ultrasonication on droplet size in biodiesel mixtures. *Journal of the American Oil Chemist's Society*, Vol.84, pp. 877-884.

Xi, Y.Z., Davis, R.J. (2008) Influence of water on the activity and stability of activated Mg–Al hydrotalcites for the transesterification of tributyrin with methanol. *Journal of Catalysis*, Vol.254, pp. 190-197.

Xie, W.L., Peng, H., Chen, L.G. (2006) Calcined Mg-Al hydrotalcites as solid base catalysts for methanolysis of soybean oil. *Journal of Molecular Catalysis A: Chemical*, Vol.246, pp.24-32.

Xie, W.L., Peng, H., Chen, L.G. (2006) Transesterification of soybean oil catalyzed by potassium loaded on alumina as a solid-base catalyst. *Applied Catalysis A: General*, Vol.300, pp.67-74.

Xu, L.L., Li, W., Hu, J.L., Yang, X., Guo, Y.H. (2009) Biodiesel production from soybean oil catalyzed by multifunctionalized Ta_2O_5/SiO_2-$[H_3PW_{12}O_{40}/R]$ (R = Me or Ph) hybrid catalyst. *Applied Catalysis B: Environmental*, Vol.90, pp.587-594.

Zięba, A., Drelinkiewicz, A., Chmielarz, P., Matachowski, L., Stejskal, J. (2010) Transesterification of triacetin with methanol on various solid acid catalysts: A role of catalyst properties. *Applied Catalysis A: General*, Vol.387, pp.13-25.

Zong, M.H., Duan, Z.Q., Lou, W.Y., Smith, T.J., Wu, H. (2007) Preparation of a sugar catalyst and its use for highly efficient production of biodiesel. *Green Chemistry*, Vol.7, pp.434-437.

8

Heterogeneous Catalysts Based on $H_3PW_{12}O_{40}$ Heteropolyacid for Free Fatty Acids Esterification

Marcio Jose da Silva[1], Abiney Lemos Cardoso[1], Fernanda de Lima Menezes[1],
Aline Mendes de Andrade[1] and Manuel Gonzalo Hernandez Terrones[2]
[1]Federal University of Viçosa/Chemistry Department,
[2]Federal University of Uberlândia/Chemistry Institute,
Brazil

1. Introduction

1.1 Biodiesel chemical background

The inevitable exhaustion of the fossil diesel reserves, besides the environmental impact generated by the green-house effect gas emission by these fuels has provoked the search by renewable feedstokes for energy production (Srivastava & Prasad, 2000; Sakay et al., 2009). Due to this crescent demand, the industry chemistry in all parts of world has search to develop environment friendly technologies for the production of alternative fuels (Di Serio et al., 2008; Marchetti et al., 2007). Biodiesel is a "green" alternative fuel that has arisen as an attractive option, mainly because it is less pollutant than its counterpart fossil and can be obtained from renewable sources (Maa & Hanna, 1999).

Although it is undeniable that biodiesel is a more environmentally benign fuel, its actual production process cannot be classified as "green chemistry process" (Kulkarni et al., 2006). The major of the biodiesel manufacture processes are carry out under alkaline or acid homogeneous catalysis conditions, where is not possible the recycling catalyst, resulting in a greater generation of effluents and salts from neutralization steps of the products and wastes (Kawashima et al., 2008). Moreover, there are some important points related to raw materials commonly used, such as high costs, besides to crescent requirements of large land reserves for its cultivation.

1.2 Production of biodiesel from triglycerides transesterification reactions

Currently, the biodiesel is manufactured from alkaline transesterification of edible or non-edible vegetable oils via a well-established industrial process (Maa & Hanna, 1999). The transesterification reaction proceeds well in the presence of some homogeneous catalysts such as alkaline metal hydroxides and Brønsted acids (Demirbas, 2003). Traditionally, sulfuric acid, hydrochloric acid, and sulfonic acid are usually preferred as acid catalysts. (Haas, 2005). The catalyst is dissolved into alcohol (methanol or ethanol) by vigorous stirring in a reactor. The vegetal oil is transferred into the biodiesel reactor and then the catalyst/alcohol mixture is pumped into the oil (Demirbas, 2003). However, the use them usually require drastic reaction conditions, i.e., high temperature and elevated pressure

(Lotero et al., 2005). In addition, serious drawbacks related to its conventional production have aroused a special attention to biodiesel industry. Some of the natural oils or animal fats contain considerable amounts of free fatty acids (FFA), which are undesirable for the transesterification processes. These important features have hardly affected the final cost to biodiesel production (Haas, 2005).

1.3 Production of biodiesel from FFA esterification reactions

An attractive alternative for lower biodiesel price is produce it directly from domestic reject such as used cocking oil and waterwastes generated by food industry (Lou et al., 2008). Nevertheless, since these low cost lipidic feedstokes are rich in FFA, it's conversion into biodiesel is not compatible with alkaline catalysts. Nevertheless, different approaches have been proposed to get rid of this problem, and frequently, two alternative pathways have been employed for produces biodiesel from these kinds of resources. At first, a two-stage process that requires an initial acid-catalyzed esterification of the FFA followed by a base-catalyzed transesterification of the triglycerides; and secondly, a single-process that makes exclusive use of acid catalysts that promote both reactions simultaneously (Dussadee et al., 2010; Zullaikah et al., 2005).

Nowadays, the catalysts conventionally used in the FFA esterification reactions are Brønsted acids and work in a homogeneous phase (Lotero et al., 2005). Acids can catalyze the reaction by donating a proton to the FFA carbonyl group, thus making it more reactive. It should be mentioned that even though traditional mineral acids catalysts are an inexpensive catalysts able to those processes, they are highly corrosive, are not reusable, and results in a large generation of acid effluents which should be neutralized leaving greater amount of salts and residues to be disposed off into environment (Di Serio, 2007). Indeed, the reduction of environmentally unacceptable wastes is a key factor for developing less pollutants and advanced catalytic processes (Haas, 2005).

Thus, to develop alternative catalysts for the direct conversion into biodiesel of lipid wastes which are basically constituted of FFA, or yet for the pre-esterification of feedstokes that has high acidity seem be also a challenge to be overcome (Demirbas, 2008). Lewis acids can be interesting alternative catalysts for biodiesel production (Corma & Garcia, 2003). Nevertheless, their high cost, the manipulation difficult and the intolerance to water of compounds traditionally used such as BF_3 and others common reagents of organic synthesis, also does not favor the use of these later in FFA esterification at industrial scale (Di Serio et al., 2005).

For all these reasons, to develop recyclable alternative catalysts for FFA esterification presents on inexpensive raw materials and food industry rejects can be an option strategically important, and undoubtedly can make the biodiesel with more competitive price using a cleaner technology (Lotero at al., 2005).

1.4 Lewis or Brønsted acids heterogeneous catalysts for biodiesel production

Recent advance in heterogeneous catalysis for biodiesel production has the potential to offer some relief to the biodiesel industry by improving its ability to process alternative cheaper raw material, and to use a shortened and low cost manufacture process. Even though many alkaline heterogeneous catalysts have been reported as highly active for biodiesel synthesis, they still cannot tolerate acidic oils with FFA content 3.5%, which are frequently used as raw material (DiMaggio et al., 2010). Contrarily, solid acids catalysts are more tolerant to FFA and are potentially less corrosive for the reactors. Consequently, these catalysts have been increasingly used in biodiesel production processes (Hattori, 2010).

A plethora of works have described the development of heterogeneous catalysts based on acids solids, which appear to offer an attractive perspective to turn the biodiesel production more environment friendly (Kiss et al., 2006; Jothiramalingam & Wang, 2009; Refaat, 2011). These solid catalysts, which normally present Lewis acidity, are easily separated from the reaction medium and are potentially less corrosive for the reactors. Normally, these processes focus on transesterification reactions of the triglycerides presents in the vegetable oils, which after react with methanol are converted into biodiesel. However, serious technological drawbacks such as drastic conditions reaction, the strict control of raw material quality in relation to water content, beyond of the leaching catalyst provoked by presence of alcohol besides water generated into reaction medium seems suggest that those process yet are hard to become effective (Kozhevnikov, 2009).

Particularly, the authors have concentrating efforts in developing alternative processes of esterification based on two recyclable catalysts linked to both acid types:

i.   heteropolyacids, with a special highlighted for the dodecatungstophosphoric acid (H$_3$PW$_{12}$O$_{40}$12H$_2$O) (Silva et al., 2010; Cardoso et al., 2008);

ii. tin chloride, an simple, easily handling, water tolerant and inexpensive Lewis acid (Cardoso et al., 2009; da Silva et al., 2010).

On the hand, catalysis by heteropolyacids of the Keggin's structure such as H$_3$PW$_{12}$O$_{40}$ is one of the most important and growing areas of research in recent years (Timofeeva, 2003). They have been extensively used in both homogeneous and heterogeneous catalysis (Misono et al, 2000; Sharma et al., 2011).

On the other hand, the use SnCl$_2$ catalyst is also most attractive, because it is solid, commercially available, and easy to handle. Moreover, its display remarkably tolerance to water, has an economically cost effective, and can be used in recyclable processes (Cardoso et al., 2008). Herein, the authors investigate the catalytic activity of heterogeneous catalysts based on acid solids composites (e.g. H$_3$PW$_{12}$O$_{40}$ supported on silicon, niobium and zirconium oxides) towards the esterification of oleic acid with ethanol.

1.5 Keggin heteropolyacid catalysts: a brief introduction

Tungtstophosphoric acid (H$_3$PW$_{12}$O$_{40}$) is a heteropolyacid largely used, in special under heterogeneous catalysis conditions. As a homogeneous catalyst the H$_3$PW$_{12}$O$_{40}$ has showed higher activity, selectivity and safety in handling in comparison to conventional mineral acids (Cardoso et al., 2008). Recent works have shown that the Keggin-type H$_3$PW$_{12}$O$_{40}$, for which the physicochemical and catalytic properties have been fully described, is an efficient super-acid that can be used in homogeneous or heterogeneous phase (Kozhevnikov, 1998). Moreover, in the heterogeneous phase, supported on several solid matrixes, heteropolyacid composites also have showed highly efficient as catalysts in several types of reactions (Pizzio et al., 1998; Timofeeva et al., 2003; Sepulveda et al., 2005).

The activity of H$_3$PW$_{12}$O$_{40}$ catalyst supported on zirconia was assessed in transesterification reactions with methanol (Sunita et al., 2008); high yields FAMEs were achieved in reactions performed at temperatures of 200 °C. On the other hand, impregnated H$_3$PW$_{12}$O$_{40}$ heteropolyacid on four different supports (i.e. hydrous zirconia, silica, alumina, and activated carbon) also were investigated and converting low quality canola oil containing to biodiesel at 200 °C temperature (Kulkarni et al., 2006). Recently, the use of an impregnation route to support H$_3$PW$_{12}$O$_{40}$ on zirconia in acidic aqueous solution and further applied in the oleic acid esterification with ethanol was described (Oliveira et al., 2010). Those authors verified that 20% w/w H$_3$PW$_{12}$O$_{40}$/ZrO$_2$ was the most active catalyst (*ca.* 88% conversion,

4 h reaction, with 1:6 FA:ethanol molar ratio and 10% w/w of the catalyst in relation to FA. However, a minor leaching of catalyst (*ca.* 8% w/w related to the initial loading), affected drastically its efficiency, resulting in decreases yielding obtained from its reuse.

2. Results and discussion

2.1 General aspects

Herein the $H_3PW_{12}O_{40}$ catalyst were supported on three different solid matrixes (i.e. silicon, niobium, and zirconium oxides) by impregnation in ethanol solutions under different loads (*ca.* 10, 30 and 50% w/w). The solids were characterized by FTIR spectroscopy and the $H_3PW_{12}O_{40}$ catalyst content was determined by UV-Vis and AAS spectroscopy analysis.

2.2 Syntheses of the $H_3PW_{12}O_{40}$ catalysts

Differently than others supports, which were used as received, zirconium oxide was obtained from thermal treatment of $ZrOCl_2.8H_2O$ salt at 300 °C during 4 hours. Composites of $H_3PW_{12}O_{40}$ supported on silicon, niobium and zirconium oxides were prepared via impregnation method (Pizzio et al., 1998). During preparation, ethanol solutions of $H_3PW_{12}O_{40}$ in hydrochloric acid 0.01 mol L^{-1} were used to avoid any hydrolysis. All composites were prepared with concentrations depending upon the loading required to the support (e.g. 10, 30 and 50% w/w $H_3PW_{12}O_{40}$) using 10 ml of the solution per gram of support. The addition of the support to the solution formed a suspension, which after stirred, was evaporated at 80 °C until dryness. All samples of supported heteropolyacid were dried at 100 °C for 12 h and then thermally treated for 4 h at 200 or 300 °C in air.

2.3 FTIR spectra of the supported heteropolyacid catalysts: $H_3PW_{12}O_{40}/SiO_2$, $H_3PW_{12}O_{40}/Nb_2O_5$ and $H_3PW_{12}O_{40}/ZrO_2$

The supported $H_3PW_{12}O_{40}$ composites were analyzed by FTIR aims to confirm the presence of the Keggin anion structure on support employed. The $PW_{12}O_{40}^{3-}$ Keggin ion structure is well known, and consists of a PO_4 tetrahedron surround by four W_3O_{13} groups formed by edge-sharing octahedral (Pope, 1983). These groups are bonded each other by corner-sharing oxygens. This structure gives rise to four types of oxygen atoms, being responsible for the fingerprint bands of the $PW_{12}O_{40}^{3-}$ Keggin ion (*ca.* 1200 - 700 cm^{-1}). FTIR spectra were obtained from all samples with different content of HPW (*ca.* 10, 30 and 50% w/w). However, the typical bands of the Keggin ions were more evident for samples with HPW contents of 30 and 50 % w/w. Herein, only the FTIR spectra of the composites with 30 % w/w $H_3PW_{12}O_{40}$, which were thermally treated at temperature of 100, 200 and 300 °C are shown. Figures 1-3 shows the characteristic bands for absorptions of v (P-O) and v (W-O) bonds existent on $H_3PW_{12}O_{40}$ composites. All FTIR spectra of both supported $H_3PW_{12}O_{40}$ catalyst or pure are displayed in Figures 1-3.

When niobium oxide was the support, only a stronger band at 1080 cm^{-1} relative to v (P-O) bond was easily observed (Figure 1). All others bands were overlapping by support bands. Conversely, when the support employed was the SiO_2, all the bands related to others oxygen atoms were observed v (W = $O_{tethraedric}$) bond at 985 cm^{-1}; v (W-O_{cubic}-W) bond, at 895 cm^{-1}, and v (W-O-W) bond, at 795 cm^{-1}; only the band of v (P-O) bond was not visible.

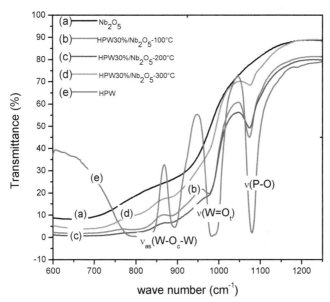

Fig. 1. FTIR spectra of (30% w/w HPW) H₃PW₁₂O₄₀ composites (a) Nb₂O₅; (b) HPW 30%/ Nb₂O₅-100°C; (c) HPW 30% Nb₂O₅-200°C; (d) HPW 30%/ Nb₂O₅-300°C; (e) HPW

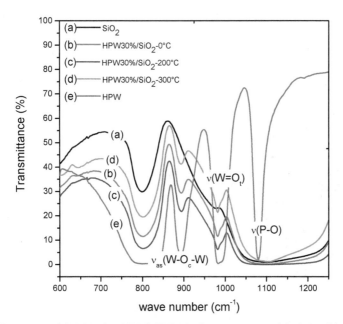

Fig. 2. FTIR spectra of (30 %w/w HPW) H₃PW₁₂O₄₀ composites (a)- SiO₂; (b)- HPW 30%/SiO₂-100°C; (c) HPW 30%/SiO₂-200°C; (d) HPW 30%/SiO₂-300°C; (e)- HPW.

These bands are preserved on the silicon-supported catalyst samples, but they are slightly broadened and partly obscured because of the strong absorptions of silica at 1100 and 800 cm^{-1} region.

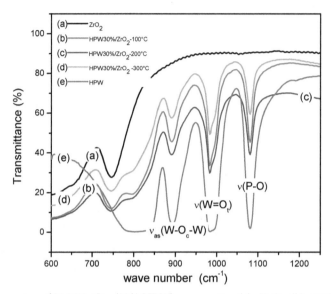

Fig. 3. FTIR spectra of $H_3PW_{12}O_{40}$ (30% HPW) composites (a)- ZrO_2; (b)- HPW/ZrO_2-100 °C (c) HPW 30%/ZrO_2-200°C; (d) HPW 30%/ZrO_2-300°C; (e)- HPW.

In Figure 3, where FTIR spectra obtained from HPW composites supported on ZrO_2 are shown, all characteristics bands of the Keggin anion are present.

In general, FTIR spectra of the HPW composites on different supports were not affected by temperature of thermal treatment. On the temperature range studied herein, all they have shown similar characteristics. However, a measured of interaction strength of HPW with support may be obtained from shift of more well defined bands to a region of lower wave number in comparison with the same band present on HPW pure (Figures 1-3).

2.4 UV-Vis spectra of the supported heteropolyacid catalysts: $H_3PW_{12}O_{40}/SiO_2$, $H_3PW_{12}O_{40}/Nb_2O_5$ and $H_3PW_{12}O_{40}/ZrO_2$

Beckman DU-650 UV-Vis spectrophotometer and quartz cells of 1.0 and 0.1 cm pathlength were employed for the adsorption experiment and measurements of $H_3PW_{12}O_{40}$ spectra, respectively (Oliveira et al., 2010). The concentration of $H_3PW_{12}O_{40}$ on catalysts was measured by UV-Vis spectroscopy before and after 6 hours of adsorption. The content of HPW in the solid was determined by AAS. In all composites yielding upper of 95% of impregnation were achieved.

2.5 Catalytic tests
2.5.1 Reaction conditions

The reactions conditions used were based on typical heterogeneous process (Figure 4). The catalyst is recovered from solution from simples filtration; the ethanol used in excess is dried and reused in other catalytic run, similarly to solid catalyst. As will show on next section, ethanol in excess not favors the ester formation under these reaction conditions.

The load catalyst used when the composites have 50% w/w HPW is corresponding to *ca.* 1 mol % in relation to oleic acid; in all reactions 1 mmol of oleic acid is used against 0.0087 mmols of HPW present in 50 mg of catalyst.

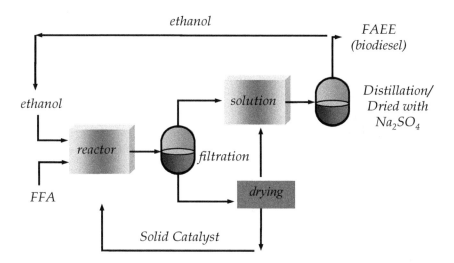

Fig. 4. Scheme of a typical acid solid-catalyzed process of FFA esterification in liquid phase

2.5.2 The effect of support on catalytic activity of HPW composites

The low surface area of solid $H_3PW_{12}O_{40}$, which implies a small amount of H⁺ ions available on the surface; for circumvent these problems, three supports with a higher surface area were selected on this study. When solid supported heterogeneous catalyst are prepared, important aspects such as temperature of the thermal treatment, method of synthesis, type and precursor nature and also of the support, besides catalyst loading can affect drastically the efficiency of catalyst (Hattori, 2010).

Herein the temperature of thermal treatment was the parameter selected for an adequate comparison between the catalytic activities of different HPW composites. High temperatures may favor the reduction of support surface area (300 °C) and lower temperatures (100 °C) may favor catalyst leaching when impregnation is synthesis method; for these reasons, the authors selected results obtained with catalyst treated at 200 °C as displayed in Figure 5.

However, another important aspect that can be affected by thermal treatment is the water content on both support and HPW catalyst. All solid supports were completely dried (*ca.* 120 °C) before of the HPW composite synthesis. Conversely, termogravimetry analysis results described in literature (Essayem et al., 1999) revealed that for the zirconium containing HPW, the loss of crystallization water upon the thermal treatment at 120 °C

which retains six water mols per mols Keggin ion. After activation at 200 °C, HPW still
retains some crystallization water molecules. (Morim et al., 2007).

The thermal treatment herein employed was the same for all supported-composite; so is
reasonable conclude that although not quantitatively determined, water effect act equally
onto both composites.

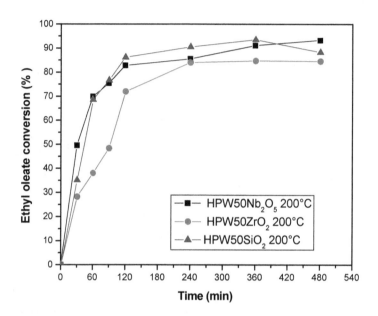

Reaction conditions: oleic acid (1.0 mmol); ethanol (155.0 mmols); catalyst (50.0 mg); 60 °C

Fig. 5. Oleic acid esterification with ethanol catalyzed by HPW 50% w/w composites
supported on niobium, zirconium and silicon treated at 200 °C temperature

The HPW 50% w/w/niobium composite is strongest Lewis acid support; nevertheless,
Figure 5 reveals the all catalyst have a very close behavior. The HPW/Nb₂O₅ composite was
the catalyst selected to assess the effects of others reaction parameters because there are
scarce data on literature; moreover, as will showed it was the catalyst more efficient and less
leached in reactions. All results obtained on HPW/niobium-catalyzed oleic acid
esterification with ethanol are highlighted in next sections.

2.5.3 Temperature effects of the thermal treatment on catalytic activity of the HPW/niobium composites

The esterification of oleic acid with ethanol conducted in the absence of the acidic catalysts
(HPW) produced no significance yields of the corresponding ethyl oleate in spite of the high
molar ratio of ethanol/oleic acid used. For instance, only a very low oleic acid to ethyl oleate
conversion (*ca.* 10%) was achieved even after a reaction time as long as 8 h (Figure 6).
Moreover, despite Lewis acidity of the support, when in presence only of niobium, a poor
conversion of oleic acid into ethyl oleate was also reached (Figure 6).

Conversely, in the presence of $H_3PW_{12}O_{40}$ pure or niobium-supported and after a reaction time of 8 h much greater yields (*ca.* 86%) were attained, as concisely displayed in Figure 5. In all reactions, a high selectivity for the ethyl oleate greater than 90 % (analysis) was achieved, determined by GC analyses (no showed herein). Investigating the performance of supported HPW can be observed that the best and worst results were obtained when the HPW/niobium composites were treated at 100 and 300 °C temperatures. A possible leaching of catalyst (see next section) and the reduction of surface area provoked by high temperature of thermal treatment may be reasonable explanations.

On the other hand, the highest conversion was obtained when a mechanic mixture of niobium and $H_3PW_{12}O_{40}$ was used, probably due the simultaneous presence of the first and second catalyst; this later soluble and consequently more reactive (Lewis and Brønsted acids respectively).

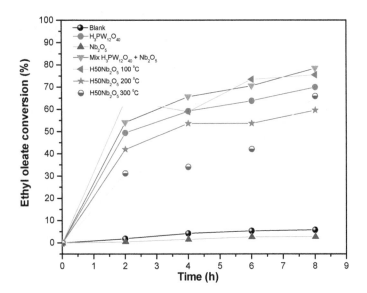

Reaction conditions: catalyst (50.0 mg); oleic acid (1.0 mmol); ethanol (155.0 mmols); 60°C.

Fig. 6. $H_3PW_{12}O_{40}/Nb_2O_5$-catalyzed oleic acid esterification with ethanol

2.5.4 The effect of HPW loading on catalytic activity of the HPW/niobium composites

In many cases, there are obvious approaches to improving and optimizing the yielding of catalytic reactions. Among the mains, is highlighted an increase on amount of reactants and of the catalyst. Recognized, the catalyst load can affect remarkably the efficiency of catalyst. Kinetic curves obtained from HPW/niobium-catalyzed esterification reactions with loads of HPW equal to 10, 30 and 50 % w/w respectively are shown in Figures 7-9. Because the temperature used on the thermal treatment may also affect both stability and activity of catalyst, three results obtained at three different temperatures are reported.

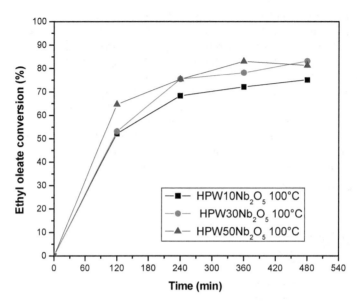

Reaction conditions: catalyst (50.0 mg); oleic acid (1.0 mmol); ethanol (155.0 mmols); 60°C.

Fig. 7. Effect of the HPW load on HPW/Nb$_2$O$_5$-100 °C-catalyzed oleic acid esterification with ethanol

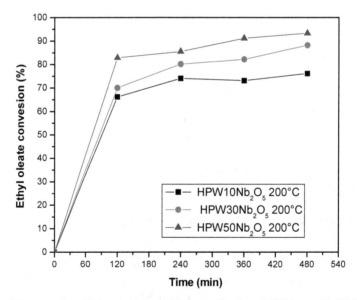

Reaction conditions: catalyst (50.0 mg); oleic acid (1.0 mmol); ethanol (155.0 mmols); 60°C.

Fig. 8. Effect of the HPW load on HPW/Nb$_2$O$_5$-200 °C-catalyzed oleic acid esterification with ethanol

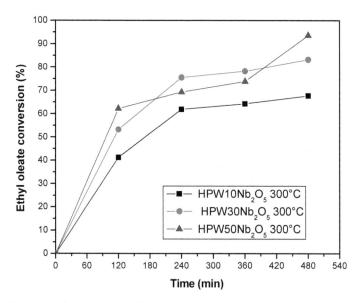

Reaction conditions: catalyst (50.0 mg); oleic acid (1.0 mmol); ethanol (155.0 mmols); 60°C.

Fig. 9. Effect of the HPW load on HPW/Nb₂O₅-300 °C-catalyzed oleic acid esterification with ethanol

Although literature data described that occur a significance decreases on surface area with an increase of acid content, which may then reduce its catalytic activity (Dias et al., 2003), results displayed in Figures 6 to 8 suggest that a higher HPW load increases the efficiency of HPW/Nb₂O₅ catalyst. Interestingly, it's occurred independently of the thermal treatment employed on synthesis of these catalysts (Figures 7-9).

2.5.5 Evaluating catalyst leaching

Leaching affects the industrial application as extensive leaching may threaten the reusability and the environmental sustainability of catalyst (Di Serio et al., 2010). Conceptually, catalyst leaching is usually associated with a phase boundary. For example, the active component of an insoluble acid solid catalyst might slowly leach into solution by some mechanism, perhaps involving bond breaking. When the catalyst has leached into a product phase, the sample should exhibit some catalytic activity. Thus, an efficient procedure that allows evaluates if there is any leaching is remove the catalyst out of the reaction and continue to run in your absence. Figures 10 to 12 displayed kinetic curves of reactions catalyzed by HPW/niobium composite before and after its remove.

It was found that the composites obtained at temperatures of 200 or 300 °C, seems be more stable under reactions conditions; noticeably, after catalyst remove the conversion of oleic acid into ethyl oleate remains constant. However, when the catalyst was synthesized at 100 °C, there was an increase in the conversion of oleic acid, suggesting that possibly a part of HPW can has been lixiviated to reaction solution. Interesting, the same occurred for the catalyst supported on zirconium and silicon (Figures 13 and 14).

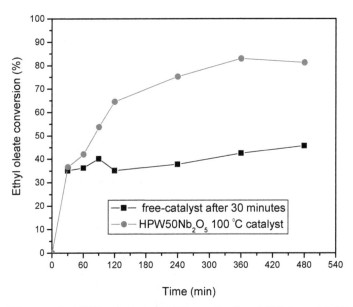

Reaction conditions: catalyst (50.0 mg); oleic acid (1.0 mmol); ethanol (155.0 mmols); 60°C.

Fig. 10. Effect of the HPW leaching on HPW/Nb₂O₅-100 °C-catalyzed oleic acid esterification with ethanol

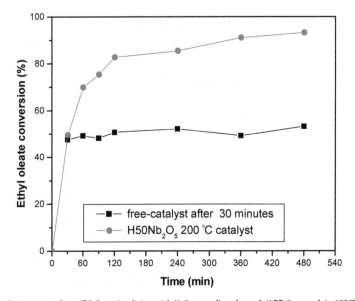

Reaction conditions: catalyst (50.0 mg); oleic acid (1.0 mmol); ethanol (155.0 mmols); 60°C.

Fig. 11. Effect of the HPW leaching on HPW/Nb₂O₅-200 °C-catalyzed oleic acid esterification with ethanol

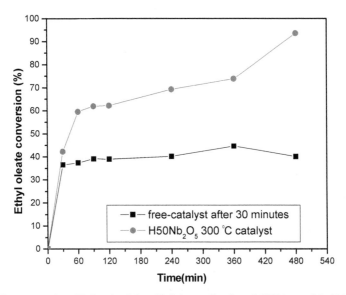

Reaction conditions: catalyst (50.0 mg); oleic acid (1.0 mmol); ethanol (155.0 mmols); 60 °C.

Fig. 12. Effect of the HPW leaching on HPW/Nb_2O_5-300 °C-catalyzed oleic acid esterification with ethanol

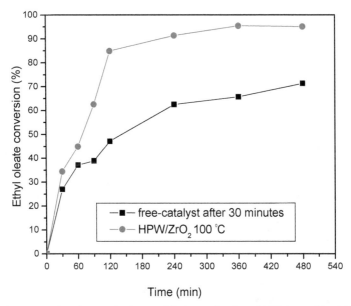

Reaction conditions: catalyst (50.0 mg); oleic acid (1.0 mmol); ethanol (155.0 mmols); 60 °C

Fig. 13. Effect of the HPW leaching on HPW/ZrO_2-100 °C-catalyzed oleic acid esterification with ethanol

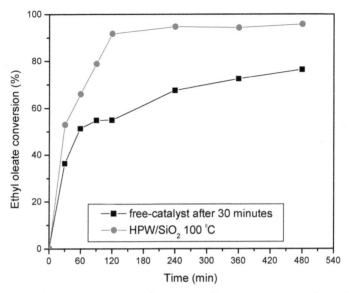

Reaction conditions: catalyst (50.0 mg); oleic acid (1.0 mmol); ethanol (155.0 mmols); 60 °C

Fig. 14. Effect of the HPW leaching on HPW/SiO$_2$-100 °C-catalyzed oleic acid esterification with ethanol

Various measures of catalyst leaching must be interpreted based in others contexts. For example, atomic absorption spectroscopy and ICP–MS are very sensitive analytical methods. However, a simple qualitative procedure can be used based only on visual observation; the addition of ascorbic acid to a solution containing HPW soluble assume blue color. Herein, its procedure allows easily confirm the catalyst leaching treated at 100 °C temperature; contrarily, in the runs with HPW/niobium-200 °C catalyst the solution remained with color unaltered (pale yellow).

2.5.6 Recovery and reuse of catalyst
The greatest advantage of the heterogeneous goal of this study over the homogeneous catalyst is the prolonged lifetime of the solid catalyst for ethyl esters production. However, leaching of catalyst components can cause its deactivation quickly. Herein, the stability of HPW 50 % w/w/niobium-200 °C after successive protocols of recovery/reuse was assessed (Figure 15). The recovery yields of solid catalyst isolated from procedure of filtration are most commonly determined gravimetrically.

A remarkable result was observed as the HPW/niobium catalytic activity stayed almost unaltered even after three recovery/reutilization cycles. However, it should be noted that a weights of catalyst fresh (ca. 20% in relation to started weight).

It was found that although recovery rate has been kept constant (*ca*. 72-75 %) in all catalytic runs, its suggest that the catalyst leach to solution; however, in Figures 10 to 12 it was demonstrated that oleic acid conversion remains unaltered after catalyst remove. This observation suggests an absence of leaching of catalyst. Probably, the procedure used is not efficient as desired.

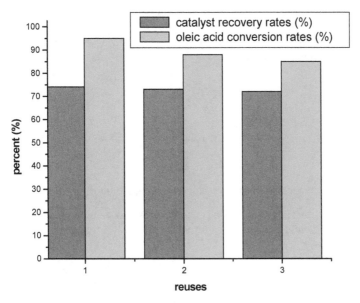

^aReaction conditions: catalyst (50.0 mg); oleic acid (1.0 mmol); ethanol (155.0 mmols); 60 °C
^bRates recovery calculated from initial catalyst mass
^cIn all runs fresh catalyst was added to reaches 50.0 mg mass

Fig. 15. Recovery Yields of HPW 50% w/w/niobium catalyst obtained by the filtration[a,b,c] procedures and oleic acid conversion rates obtained from its esterification with ethanol

The procedure employed for catalyst recovery involves its separation from reaction by filtration, washed with ethyl ether and drying at 100 °C; then the catalyst has its mass determined. Losses of mass through of these several steps may be occurring. A more detailed treatment of the recovery procedure of catalyst may lead to efficient methods, which can reaches higher recovery rates. The authors are developing studies on this direction.

2.5.7 Mechanistic insights
Tunstophosphoric acid ($H_3PW_{12}O_{40}$) is strongest heteropolyacid of Keggin series being completely ionizable in water. Measurements of pKa in organic solvents showed that it is 100 units of pka more acid than sulfuric acid (Kozhevnikov, 1998); therefore is almost probably that its ionization in ethanol occur in greater extension. Thus, is possible that HPW/niobium catalyst undergoes at least a partial ionization along oleic acid esterification reaction in ethanol as described on equilibrium displayed in Figure 16.

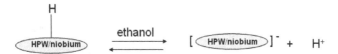

Fig. 16. Partial ionization equilibrium of HPW/niobium catalyst in ethanol solution

Consequently, if this part of the reaction pathway is similar to homogeneous systems, the others steps commonly involved in esterification reactions (e.g. protonation carbonyl group FA, attack of the alcohol molecule on protonated FA, water elimination, etc) may then proceed as described in Figure 16.

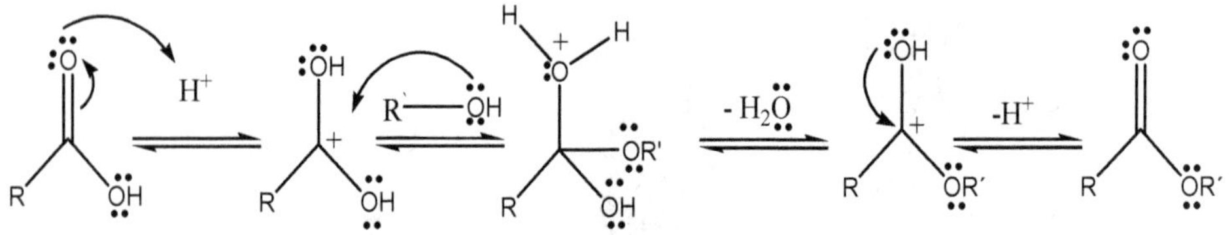

Fig. 17. Mechanism of formation ester catalyzed by free H^+ ion in solution

Conversely, is also possible that other FA molecules can be activated via protonation on surface of supported-catalyst. Thus, an alternative proposal is displayed in Figure 18.

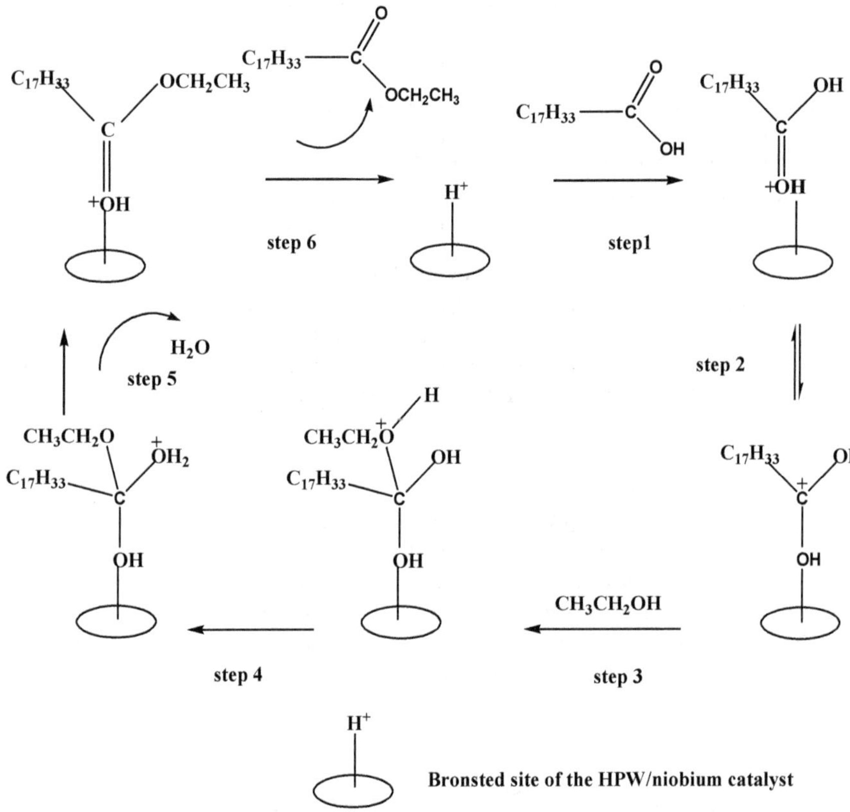

Fig. 18. Proposal of an alternative mechanism of formation ester catalyzed by free H^+ ion in the solution

In according with this mechanism (Figure 18), all steps of oleic acid esterification reaction with ethanol occur on surface of HPW/niobium catalyst. Nevertheless, is also possible that ethyl oleate formation may occur by both pathways of reaction Although both proposal are plausible, is important to note that studies in situ are require for a better and more detailed description of these mechanism of this reaction.

3. Conclusion

The efficiency of tungstophosphoric acid ($H_3PW_{12}O_{40}$) immobilized by impregnation method on silicon, zirconium and niobium oxides was assessed in the esterification of oleic acid with ethanol, at 60 °C temperature. As a general tendency, it was observed that the catalytic activity decreases in the series $HPW/Nb_2O_5 > HPW/ZrO_2 > HPW/SiO_2$ with all catalyst being treated on temperature range 100 to 300 °C. Moreover, good yielding of recovery of HPW 50% w/w/Nb_2O_5 catalyst (*ca.* 75 %) and high conversions of acid oleic were obtained in recycle experiments. From leaching tests and of the rates of recovery may be concluded that the HPW/Nb_2O_5 catalysts are stable under reaction conditions used; however the recovery procedure employed it should be enhanced. Thus, it can be concluded that although yet non-finished, present methodology offers several advantages such as high yields, simple procedure for recovery and reuse of catalyst and mild reaction conditions.
The authors hope that with this work a significant advance on the field of recoverable catalysts can has been proved.

4. Acknowledgment

The Federal University of Viçosa, Federal University of Uberlândia and Arthur Bernardes Foundation are warmly thanked for financial support. Moreover, CAPES, CNPq and FAPEMIG deserves our special thanks.

5. References

Anton, A. K.; Alexandre, C. D.; & Gadi, R. (2006) Solid Acid Catalysts for Biodiesel Production - Towards Sustainable Energy. *Advances Synthesis Catalysis*, Vol.348, pp. 75 – 81

Ayhan D.; Biodiesel A Realistic Fuel Alternative for Diesel Engines; ISBN-13: 9781846289941 2008 Springer-Verlag London Limited

Ayhan, D. (2003). Biodiesel fuels from vegetable oils via catalytic and non-catalytic supercritical alcohol transesterifications and other methods: a survey. *Energy Conversion and Management*, Vol.44, (September 2007), pp. 2093–2109, ISSN 0196-8904

Cardoso, A.L.; Neves, S.C.G.; & da Silva, M.J. (2009). Kinetic Study of Alcoholysis of the Fatty Acids Catalyzed by Tin Chloride(II): An Alternative Catalyst for Biodiesel Production. *Energy Fuels*, Vol.23 (January 2009), No. 3, pp. 1718–1722, ISNN 0887-0624

Chongkhong, S.; Tongurai, C.; Chetpattananondh, P. (2009). Continuous esterification for biodiesel production from palm fatty acid distillate using economical process. *Renewable Energies*, Vol.34, (April 2009), pp.1059– 1063, ISSN 0960-1481

Di Serio, M.; M. Cozzolino, M. Giordano, R. Tesser, P. Patrono, & E. Santacesaria. (2007). From Homogeneous to Heterogeneous Catalysts in Biodiesel Production. *Industrial Engineering Chemistry Researches* Vol.46, (August de 2007), pp 6379–6384, ISNN 0088-5885

Di Serio, M.; Tesser, R.; Casale, L.; D'Angelo, A.; Trifuoggi, M. & Santacesaria, E., (2010). Heterogeneous catalysis in biodiesel production: The influence of leaching. *Topics in Cataysis*, Vol.53 (July 2010), pp.811-819 ISNN 1022-5528

Di Serio, M.; Tesser, R.; Pengmei, L.; & Santacesaria, E. (2008) Heterogeneous Catalysts for Biodiesel Production. *Energy and Fuels*, Vol.22, (December 2007), pp. 207–217, ISSN 0887-0624

Essayem, N.; Coudurier, G.; Vedrine, J.C.; Habermarcher, D.; Sommer, J. (1999). Activation of Small Alkanes by Heteropolyacids, a H/D Exchange Study: The Key Role of Hydration Water. *Journal of Cataysis*, Vol.183, (April 1999), pp. 292-299, ISSN 00219517

Haas, M. J . (2005). Improving the economics of biodiesel production through the use of low value lipids as feedstocks: vegetable oil soapstock. *Fuel Processing Technologies*, Vol.86, (June 2005), pp. 1087–1096, ISNN: 0378-3820

Hideshi H. (2010). Solid Acid Catalysts: Roles in Chemical Industries and New Concepts. *Topics in Catalysis*, Vol.53, (June 2010), pp.432–438, ISNN 1022-5528

Kawashima, A.; Matsubara, K.; Honda, K. (2008). Development of heterogeneous basecatalysts for biodiesel production. *Bioresource Technology*, Vol.99, pp. 3439–3443, ISNN 0960-8524

Kozhevnikov I.V. (1998). Catalysis by heteropoly acids and multicomponent polyoxometalates in liquid-phase reactions. *Chemical Reviews*, Vol.98, (February 1998), No.1, pp.171–198

Kozhevnikov, I.V. J. (2009) Heterogeneous acid catalysis by heteropoly acids: approaches to catalyst deactivation. *Journal of Molecular Catalysis A*, Vol.305, (june 2009), pp. 104–111, ISNN 1381-1169

Kulkarni, M. G.; Gopinath, R.; Meher, L.C.; & Dalai, A. K. (2006). Solid Acid Catalyzed Biodiesel Production by Simultaneous Esterification and Transesterification. *Green Chemistry*, Vol.8, (September 2006), pp. 1056–1062, ISSN 1463-9262

Lotero E.; Liu, Y.; Lopez, D. E.; Suwannakarn, K.; Bruce, D. A. & Goodwin, J. G. (2005) Synthesis of biodiesel via acid catalysis. *Industrial Engineering Chemistry Researches*, Vol.44, (January 2005), pp. 5353–5363, ISNN 0088-5885

Lou, W. Y.; Zong, M. H.; & Duan ,Z. Q. (2008). Efficient production of biodiesel from high free fatty acid-containing waste oils using various carbohydrate-derived solid acid catalysts. *Bioresource Technology* Vol.99, (December 2008), pp. 8752–8758, ISNN: 0960-8524

Maa, F.; & Hanna, M.A. (1999). Biodiesel production—a review. *Bioresource Technology*, Vol.70, (February 2007), pp. 1–15, ISSN 0960-8524

Makoto M.; Izumi O.; Gaku K.; & Atsushi A. (2000). Heteropolyacids. Versatile green catalysts usable in a variety of reaction media. *Pure Applied Chemistry*, Vol. 72, No. 7, pp. 1305-1311, ISSN 0033-4545

Marchetti, J. M.; Miguel, V. U.; & Errazu A. F. (2007). Possible methods for biodiesel productions. *Renewable Sustainable Energies Reviews*, Vol.11, pp.1300- 1311, ISSN 1364-0321

Morin, P. B.; Hamad, G.; Sapaly, M. G.; Carneiro, R.; Pries, P.G. O.; Gonzalez, W. A.; Sales, A. E.; Essayem, N. (2007). Transesterification of rapeseed oil with ethanol I. Catalysis with homogeneous Keggin heteropolyacids. *Applied Catalysis A: General*, Vol. 330, (October 2007), pp. 69–76, ISSN: 0926-860X

Oliveira, C. F, Dezaneti, L. M.; Garcia, F. A. C.; Macedo, J. L.; Dias, J. A.; Dias, S. C. L.; Alvim, K. S. P. (2010). Esterification of oleic acid with ethanol by 12-tungstophosphoric acid supported on zirconia. *Applied Catalysis A: General*, Vol.372, (January 2010), pp. 153–161, ISSN: 0926-860X

Patil, P.D.; & Deng. (2009) Optimization of biodiesel production from edible and non-edible vegetable oils. *Fuel*, Vol.88, pp. 1302–1306. ISSN 00162361

Pizzio, L. R.; Caceres, C. V.; Blanco, M. N. (1998). Acid catalysts prepared by impregnation of tungstophosphoric acid solutions on different supports. *Applied Catalysis A: General*, Vol.167, (February 1998,) pp.283-294, ISSN: 0926-860X

Pope, M.T. *Heteropoly and Isopoly Oxometalates*, Springer-Verlag, Berlin, 1983. ISBN 3-540-11889-6

Prasad, R; & Srivastava, A. (2000). Triglycerides-based diesel fuels. Renewable and Sustainable *Energy Reviews*, Vol.4, (July 1999), pp. 111-133, ISSN 1364-0321

Rajabathar, J.; & Ming K. W. (2009). Review of Recent Developments in Solid Acid, Base, and Enzyme Catalysts (Heterogeneous) for Biodiesel Production via Transesterification. *Industrial Engineering Chemistry Researches*, Vol.48, No.13, pp. 6162–6172, ISSN 0088-5885

Rattanaphra, D.; Harvey, A. & Srinophakun, P. (2010). Simultaneous Conversion of Triglyceride/Free Fatty Acid Mixtures into Biodiesel Using Sulfated Zirconia. *Topics in Catalysis*, Vol.53, (March 2010), pp.773–782, ISNN 1022-5528

Sakai, T.; Kawashima, A.; & Koshikawa, T. (2009). Economic assessment of batch biodiesel production processes using homogeneous and heterogeneous alkali catalysts. *Bioresource Technology* Vol.100, (July 2009), pp.3268–3276, ISSN 0960-8524

Sepulveda, J.H.; Yori, J.C.; Vera, C.R. (2005). Repeated use of supported H$_3$PW$_{12}$O$_{40}$ catalysts in the liquid phase esterification of acetic acid with butanol. *Applied Catalysis A: General*, Vol.288, (July 2005), pp.18–24, ISNN: 0926-860X

Shuli Y. C. D.; Siddharth, M.; Manhoe K.; Steven O. S.; & Simon, K. Y. N. (2010). Advancements in Heterogeneous Catalysis for Biodiesel Synthesis. *Topics in Catalysis*, Vol.53, (July 2010), pp.721–736, ISNN 1022-5528

Silva, G. W. V.; Lair, O. L.; da Silva, J. M. (2011). Novel H$_3$PW$_{12}$O$_{40}$ Catalysed Esterification Reactions of Fatty Acids at Room Temperature for Biodiesel Production. *Catalysis Letters*, Vol.135 (January 2010), pp.207-211, ISSN: 1011-372X

Silva, M. L.; Figueiredo A. P.; Cardoso, A. L.; Natalino, R.; da Silva, M. J. (2011) Effect of Water on the Ethanolysis of Waste Cooking Soybean Oil Using a Tin(II) Chloride Catalyst. *Journal American Oil Chemisty Society*, Vol. 88, (march 2011), pp.1431-1437, ISSN 0003-021X

Sunita, G; Devassy, B.M.; Vinu, A.; Sawant, D.P.; Balasubramanian, V.V.; & Halligudi, S.B. (2008). Synthesis of biodiesel over zirconia-supported isopoly and heteropoly tungstate catalysts, *Catalysis Communications*, Vol.9, (March 2008), pp.696-702, ISSN: 1566-7367

Timofeeva, M.N. (2003). Acid catalysis by heteropoly acids. *Applied Catalysis A: General*, Vol.256, (December 2003), pp.19–35, ISNN 0926-860X

Timofeeva, M.N.; Matrosova M.M.; Ilinich, G.N.; Reshetenko, T.V.; Avdeeva, L.B.; Kvon R.I,.; Chuvilin A.L.; Budneva,. A.A.; Paukshtis, E.A.; Likholobov, V.A . (2003). Esterification of n-butanol with acetic acid in the presence of $H_3PW_{12}O_{40}$ supported on mesoporous carbon materials. *Kinetic Catalysis* Vol.44, (November 2003) No.3, pp.778–787, ISSN 0023-1584

Yogesh C. S.; & Bhaskar S. (2011). Advancements in solid acid catalysts for ecofriendly and economically viable synthesis of biodiesel. *Biofuels, Bioproduts Biorefinerie,* Vol.5, (September 2010), pp. 69–92, ISSN 1932-104X

Zullaikah, S.; Lai C, C.; Vali, S. R. & Ju, Y.H. (2005). A two-step acid-catalyzed process for the production of biodiesel from rice bran oil. *Bioresource Technology,* Vol.96, (November 2005), pp. 1889–1896, ISNN: 0960-8524

Progress in Vegetable Oils Enzymatic Transesterification to Biodiesel - Case Study

Ana Aurelia Chirvase[1], Luminita Tcacenco[2],
Nicoleta Radu[1] and Irina Lupescu[3]
[1]*The National Institute for Research & Development in Chemistry and Petrochemistry,*
[2]*The National Institute for Research & Development in Biological Sciences,*
[3]*The National Institute for Research & Development in Pharmaceutical Chemistry,*
Romania

1. Introduction

These days the interest of fuels preparing from sustainable natural resources is continuously increasing due to the rising prices of the fossil fuels and the political instability in the oil producing countries. The fuels manufacturing from local vegetal resources can sustain the every country' prosperity, including rural, agricultural, economically disadvantaged regions. Nowadays only the bioethanol and the biodiesel are already produced at industrial level from sustainable raw materials.

The biodiesel is manufactured by the chemically catalysed transesterification of the triglycerides from the vegetable oils, rapeseed oil in Europe and soya oil in USA. As the methanol is often used as alcohol reagent, the reaction is consequently named methanolysis. The most applied catalysts are alkalines (especially NaOH) or mineral acids. So the biodiesel represents the methyl esters of the fatty acids from the vegetable oils. The present diesel engines can normally use a mixture of diesel with 5% v/v biodiesel.

Biodiesel contains virtually no sulfur or aromatics, and use of biodiesel in a conventional diesel engine results in substantial reduction of unburned hydrocarbons, carbon monoxide and particulate matter. The production and use of biodiesel, compared to petroleum diesel, resulted in a 78.5% reduction in carbon dioxide emissions. Moreover, biodiesel has a positive energy balance.

The chemical transesterification applied at industrial level has important advantages, but also limitations: in spite of the high conversion yields and the short reaction duration, the global transformation is energetically intensive, the glycerol recovery is difficult, the alkaline catalyst must be separated, the wastewaters are to be treated by a rather complex procedure, and both the free fatty acids and water can badly influence the reaction.

These unfavourable situations can be diminished by performing the enzymatic transesterification on conditions that: (a) the immobilised lipase used as biocatalyst must be as cheap as possible; (b) one can obtain the economic efficiency of the whole biotransformation process similar to that characteristic to the chemical process, these objectives being presented function of the research methodology and results. The comparison between the chemical way and the enzymatic way is presented in the Table 1.

Criterion	Alkaline catalysis process	Enzymatic proces
• Reaction temperature	60-70°C	30-40°C
• Free fatty acids from the vegetable oils	Saponification products	Methylic esters
• Water from the raw material	Reaction interference	No influence
• Methylic esters yield	Normal	Higher
• Glycerol recovery	Difficult	Easy
• Methylic esters purification	Repeated washing	No need
• Catalyst preparation price	Cheap	Relatively high

Table 1. Comparison between the alkaline catalysis and the enzymatic method for biodiesel preparation (Bajaj et all, 2010)

The now-a-day technological progress regarding the enzymatic transesterification is demonstrated by the realisation of 2 industrial pilots in China (Moore 2008a, 2008b; Uthoff et all, 2009) to apply this advanced methodology, though the biodiesel manufacture price still remains higher than the diesel price no matter the transesterification route, due to the raw materials high prices (Bisen et all, 2010). Developments to meet the economical framework are needed, including: (a) the introduction of the enzymatic transesterification of plant oils as a part from a comprehensive technology of complete valorisation of the vegetable oil, meaning the application of the bio refinery concept; (b) the increase of the available vegetable oil quantity with limited interference with the vegetable oils' food use; (c) the possible preparation of methanol from natural resources.

2. State of the art in the domain of biodiesel preparation by enzymatic transesterification of vegetable oils

Other advantages of using lipases in biodiesel production are: (a) ability to work in very different media which include biphasic system, and monophasic system, (b) they are robust and versatile enzymes that can be produced in bulk because of their extracellular nature in most manufacturing system, (c) when the lipase is used in a packed bed reactor, no separation is necessary after transesterification, and (d) higher thermo stability and short-chain alcohol-tolerant capabilities of lipase make it very convenient for use in biodiesel production (Ghaly et all, 2010). Until now the biodiesel manufactured by chemical catalysis is cheaper than the same product obtained by enzymatic catalysis, but in case of considering the pollution suppressing costs needed after the chemical process performing, the costs of both reaction' types could be comparable.

Enzymatic transesterification can be done with crude or purified vegetable oils, free fatty acids, residual grease from food industry or of animal origin, and residual vegetable oils from fry cooking. Beside methanol and ethanol one can also use as acyl acceptors the propanol, iso-propanol, butanol and iso-butanol. Many microorganisms, bacteria, yeasts or fungi can produce useful lipases for transesterification. Of these microorganisms, Candida antarctica, Candida rugosa, Pseudomonas cepacia, Pseudomonas fluorescens, Rhizomucor miehei, Rhizopus chinensis, Rhizopus oryzae and Thermomyces lanuginosa have produced the most effective lipases, able to perform the biotransformation with high yields. The combination of two or more lipases can increase the conversion in order to lower the cost. A combination of Candida antarctica and Thermomyces lanuginosa lipases was used to obtain a 95% conversion in methanolysis using a tert-butanol solvent. From the many lipases it is recommended to use those with reduced region specificity, but with higher substrate specificity.

The reaction can be realised either in organic solvents, or in solvent-free media (where there are only the substrates' mixture). Normally in organic solvents' systems the lipases can catalyse the biotransformation when the alcohol is added stepwise at the beginning (a „batch" system), by comparison with the free-solvent media, where the alcohol is added several times for maintaining a certain molar ratio with the oil concentration.

The **key factors affecting the enzymatic transesterification** are presented in the Figure 1.

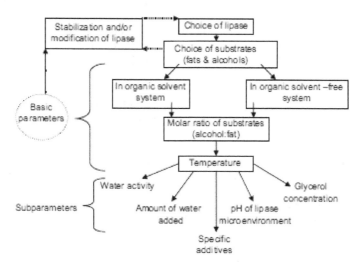

Fig. 1. Key factors of influence on the enzymatic transesterification (Antczak *et all*, 2009)

There are two categories of enzymatic biocatalysts: (1) extracellular lipases (i.e. the enzyme has previously been recovered from the cultivation broth and then purified) especially from the microbial producers *Candida rugosa*, *Candida utilis*, *Candida antarctica* and *Pseudomonas cepacia*, generally bacteria and yeasts; (2) intracellular lipases which still remain either inside or attached to the cellular wall; in both cases the enzymes are immobilized directly or together with the whole cell and this use can eliminate downstream operations and assure the enzyme recycling.

The **extracellular lipases** are mostly produced by bacteria and yeasts and the large scale production of these lipases should be economical, fast, easy and efficient. Unfortunately, the cost of specific separation and purification operations is high enough. Still the majority of immobilized lipases that are commercially available are extracellular. The most commonly used is: Novozym 435 which is the lipase from *Candida antarctica*. Meanwhile the bacteria and yeasts can probably form growth associated lipases, in a first stage, linked to the cellular membranes, then released into the cultivation medium as extracellular enzymes.

When preparing the **intracellular lipases** the costly step of purification can be eliminated and this has led to using whole cells as biocatalysts. After the intracellular production of lipases the direct use of fungal cells immobilized within porous biomass support particles as a whole biocatalyst represents an attractive process for bulk production of biodiesel (Fjerbaek *et all*, 2009)

The main criteria to choose between the two lipase types can be: (a) the bacteria and yeasts strains which biosynthesise extracellular lipases, can be considered as recommended producers based on the cultivation conditions, namely easy to apply and reproducible aerobic bioprocesses; (b) using intracellular lipases slows down the transesterification process due to mass transfer limitations.

Immobilization of an enzyme must solve both mass transfer limitations types-internal or external (last case due to formation of an external film layer). Choice of the appropriate lipase immobilization technology is determined by the following objectives: (a) long term enzyme reuse; (b) easy enzyme recovery from the reaction medium; (c) improved activity and thermal, chemical and mechanical stability of the enzyme; (d) potential to run continuous processes. The immobilization support is to be as low cost as possible, condition which is difficult to be observed when the other ones should be fulfilled at the same time (Ghaly et all, 2010). Among the great number of immobilization techniques, they can be classified under four general categories: (a) adsorption, (b) cross linking, (c) entrapment and (d) encapsulation. Adsorption seems to be the most attractive, as it is simplest and cheap, retaining high enzyme activity and allowing a good mass transfer, combined or not with the cross linking. The carriers used in adsorption via weak forces include: celite, cellulose, acrylic, silica gel, textile membranes, spherosil, sepharose, sephadex and siliconized glass. The major drawback of the adsorption is the low stability of the enzyme when adsorbed, which determines only limited reuse.

The **stability of the lipase** with low loss of the catalytic activity is the most important characteristic, when used in biodiesel preparation in connection with the enzyme **recovery and reuse.**

The most commonly used reactor type for the biodiesel enzymatic preparation is a batch-stirred tank reactor, though this biofuel must be considered as a commodity product and therefore produced in continuously operated installations. Possible alternative solutions could be packed bed reactors, fluid beds, expanding bed, recirculation membrane reactors. A wide range of configurations are applicable to perform the transesterification.

As the actual major technical limits of the enzymatic process are still the slower reaction rate by comparison with the alkaline catalysis and the risk of enzyme inactivation, with focus on process design and economy, the researchers calculate the productivity (kg biodiesel/kg enzyme) based on information from different studies and considering a range of enzyme prices from 12 to 185 USD/kg as acceptable, depending on the application characteristics, i.e. per each kg of biodiesel a biocatalyst cost of USD 0.025 could be of economic interest. An increased enzyme life of around 6 years would make enzymes competitive based on productivity again. To this must be added increased reactor costs as enzymes lead to longer space times than alkaline catalysts, but reduced separation costs and low waste water treatment costs will be the benefits.

3. Case study: Enzymatic transesterification of the rapeseed oil with yeast lipases

The chapter presents the research activity done by the authors regarding the rapeseed oil transesterification with yeast lipase, and is structured in three parts: lipase formation in aerobic bioprocessing; lipase recovery and immobilization; enzymatic transesterification with immobilized lipase produced by the yeast *Candida rugosa* DSM 70761.

3.1 Lipase formation
3.1.1 Materials and methods
Several bacteria and yeasts from own / international collections were tested for cell growth and enzyme formation, the cultivation conditions being: rotary shaker New Brunswick

Innova 40 at 300 rpm; temperature of 30ºC; Erlenmeyer flasks of 500 mL with 150 mL medium. Before their cultivation for enzyme formation the microorganisms were grown on liquid media to develop preinoculum and inoculum stages of 24 hours duration, using an inoculation volume of 5-10 % V/V. Several cultivation media, specific for the studied strains, were tested and both the cellular growth and enzymatic activity were measured.

Microorganisms and cultivation media:

Bacteria: *Pseudomonas putida (P. sp. 1) and Pseudomonas aeruginosa (P. sp. 3)*
Yeasts: *Yarrowia sp. / Candida lipolytica ATCC 8661, Candida sp. DG 8, Pseudozyma aphidis DSM 70725, and Candida rugosa DSM 70761.*

M1 for bacteria: (variant a: no rapeseed oil; variant b: with 10 mL/L rapeseed oil)	M2 for yeasts	M3 for yeasts	M4 for yeasts
Glucose: 4 g/L	Glucose: 10 g/L	KH_2PO_4: 5 g/L	Malt extract: 3.78 g/L
Peptone: 0.5 g/L	Peptone : 10 g/L	$(NH_4)_2SO_4$: 1 g/L	Peptone: 5 g/L
Yeast extract: 5 g/L	Yeast extract: 10 g/L	Yeast extract: 10 g/L	Tween 80: 4.33 g/L
Na_2SO_4: 2 g/L	NH_4Cl : 5 g/L	$MgSO_4.7H_2O$: 0.5 g/L	
KH_2PO_4: 1 g/L	Rapeseed oil: 5 g/L	Rapeseed oil: 20 g/L	Rapeseed oil: 33.7 g/L
K_2HPO_4: 3 g/L			
$MgSO_4.7H_2O$: 0.1 g/L			

Table 2. Cultivation media composition

The growth characteristics were evaluated by measuring OD_{500}; the lipase activity was determined by using the volumetric method (Tcacenco *et all*, 2010), considering one unit of lipase activity as corresponding to 1 µmol of fatty acid obtained by the hydrolysis of the triglycerides from the rapeseed / olive oil, the reaction conditions being: temperature of 37°C, pH=7, duration of 60 minutes.

Isolation of extracellular lipase was made by centrifugation (1) and ammonium sulphate precipitation (2): (1) biosynthesis medium was centrifuged at 10 000 rpm for 30 min. at 4 ºC. Clear supernatant was treated with benzamidine 2 mM and sodium azide 0.02% to prevent proteolysis and microbial attack and (2) the supernatant is precipitated with ammonium sulphate 30% at 0 ºC, then left to stand for 24 hours for achieving precipitation and centrifuged at 10 000 rpm for 30 minutes at 4ºC. The supernatant is precipitated again with 75% ammonium sulphate. After 24h, the sample is centrifuged again and the resulting product is dissolved in 8 ml TRIS buffer, pH 6.8. This crude enzyme is preserved in the freezer.

3.1.2 Results and discussion

1.   Bacteria growth and enzyme formation

The growth of both bacteria is low, only *Pseudomonas aeruginosa (P. sp.3)* grows more on the medium variant M1b, so the use of both substrates-glucose and oil seems useful. Both bacterium strains have similar small lipase activity levels, the cultivation duration of 24 h being enough for the maximum lipase production, and there is no induction by the rapeseed oil (Fig. 2).

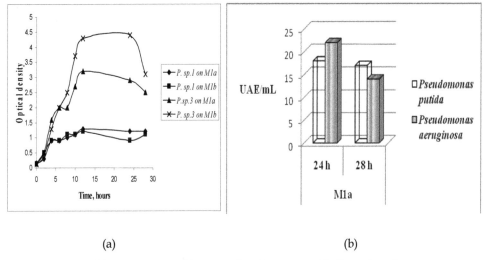

(a)   (b)

Fig. 2. Growth and lipase activity of bacteria *Pseudomonas putida* (*P. sp.1*) and *Pseudomonas aeruginosa* (*P. sp. 3*) on cultivation medium M1, a and b; (a) growth (OD); (b) lipase formation

2. Yeasts growth and lipase formation
    Experimental variants:
    A1: *Candida rugosa* DSM 70761 on M1, 48 h
    A2: *Candida rugosa* DSM 70761 on M2, 48 h
    B1: *Pseudozyma aphidis* DSM 70725 on M1, 48 h
    B2: *Pseudozyma aphidis* DSM 70725 on M2, 48 h
    C1: *Candida rugosa* DSM 70761 on M3, 48 h
    C2: *Pseudozyma aphidis* DSM 70725 on M3, 48 h
    D1: *Yarrowia (Candida lypolitica)* ATCC 8661 on M2, 24 h
    D2: *Candida sp. DG 8* on M3, 24 h

For the cultivation medium M2 the growth rate for the yeasts *Candida rugosa* DSM 70761 and *Candida lypolitica* ATCC 8661 were higher and close enough: variant D1 *Yarrowia lipolytica* with the specific growth rate of 0.2 h^{-1} ; variant A2 *Candida rugosa* with the specific growth rate of 0.15 h^{-1}. But the final enzyme activity was higher for the second yeast: *Candida rugosa* final enzymatic activity of 289.0 UAE/mL by comparison with *Yarrowia lipolytica* enzymatic activity of 106.0 UAE/mL. At the same time the growth and lipase activity of both yeasts were much higher than those of the studied bacteria. So the immobilization study was to be performed with these already mentioned yeasts. In a first step, the preliminary transesterification results, obtained by thin layer chromatography, demonstrated that both lipases have high enough catalysis activities. After the confirmation of the transesterification capacity, it was of interest to develop appropriate immobilization techniques for these lipases, so to be able to use the immobilized enzymes in several cycles of biotransformation.

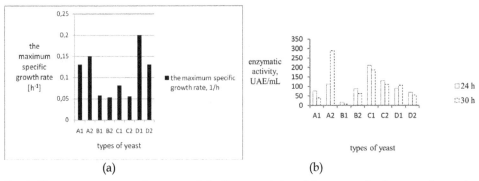

(a)   (b)

Fig. 3. The maximum specific rate and the lipase activity of the yeasts for the experimental variants A_i-D_i (i=1, 2); (a) max specific growth rate (μ^{-1}); (b) lipase activity (UAE/mL) (Chirvase *et all*, 2010)

3.2 Lipase immobilization
3.2.1 Materials and methods
The techniques by physical adsorption were chosen due to the fact they are simple and cheap, so the price of the immobilized biocatalyst is expected to be low.

a.   Lipase immobilization by adsorption on silicagel or celite support

The crude lipase obtained from *Yarrowia lipolytica* and *Candida rugosa* yeasts after the precipitation with 70% ammonium sulphate was dissolved in 0.05 M phosphate buffer, pH 7. Then the adsorbent was added until the limit activity in the supernatant is reached, respectively: for *Yarrowia lipolytica* 2.5 g silicagel G at 800 mL extract, 22 g of Celite in the same volume of extract and for *Candida rugosa* 11 g Celite at 800 ml extract. Adsorption duration was approx. 2 hours at ambient temperature and under mechanical stirring.

b.   Lipase immobilization by adsorption on chitosan support

  1.   Cross-linking with glutaraldehyde:

30 mL chitosan 1% solution was prepared by adding 2mL CH_3COOH p.a. , 19.8 mL 0.5 N NaOH by heating to 50 °C and stirring for 10 minutes to complete dissolution of chitosan. 0.5 mL 25% of glutaraldehyde was added dropwise under high stirring. Microspheres thus obtained were filtered and washed with H_2O dist. and 0.05 M phosphate buffer, pH 7. 1g wet chitosan microspheres were used for immobilization; they were suspended in 2 mL 0.05 M phosphate buffer, pH 7 and mixed with 2mL solution of lipase (*Candida rugosa*) obtained by solving the crude enzyme precipitated with ammonium sulphate into 0.05 M phosphate buffer, pH 7, 1:5 (w / v) ratio. The mixture was stirred for 1 hour at 37 °C.

  2.   Cross-linking with carbodiimide:

1g wet chitosan particles was obtained by injecting 25 mL solution of 3% chitosan into 250 mL solution of NaOH 1N and C_2H_5OH 26%. The chitosan particles were suspended into 3 mL 0.75% carbodiimide solution, prepared in 0.05 M phosphate buffer, pH 6, 25 °C. After 10 minutes of activation, the particles were washed with distilled water and transferred to 10 mL 1% lipase solution immersed in 0.05 M phosphate buffer, pH 6. The adsorption duration was 60 minutes; then the immobilized enzyme was washed 3 times with distilled water.

  3.   Cross-linking with glutaraldehyde and reduction with sodium borohydride

A mixture was prepared from 0.5 g chitosan, 1.041 mL 2M acetic acid, 25 mL distilled water and 1.041 mL of 1M sodium acetate, maintained on water bath at 50°C with stirring. For the immobilization of *Aspergillus niger* lyophilized lipase (Fluka), 0.1 g of lipase immersed in 0.5 M phosphate buffer, pH 5.6 was added to this mixture. Then 2.5 mL 50% glutaraldehyde dissolved in 25 mL double distilled water was added. The mixture rested for 30 minutes at 4 ⁰C. 0.25 g sodium borohydride was added in portions, during 15 minutes, with ice pieces to low the temperature, and finally the mixture was filtrated in vacuum. The immobilized product thus obtained was washed with double distilled water and 0.5 M phosphate buffer, pH 5.6. Lipase activity and immobilization yield were evaluated for each application.

3.2.2 Results and discussion

The final activities and isolation yields obtained when the crude lipases were separated from the cultivation medium by precipitation with ammonium sulphate are presented in the Table 3. High efficient lipases isolation was done by precipitation of the cultivation medium of *Yarrowia lipolytica* yeast with $(NH_4)_2SO_4$, a yield of 95% was got for both variants (24 hr. and 28 hr.), while when the same procedure was applied for *Candida rugosa* samples the isolation yields were lower: 62% for 24 hr. extract and only 29.6% for 28 hr. extract.

No.	Strain / duration of bioprocessing	Extract volume (mL)	Initial activity (UEA)	Quantity $(NH_4)_2SO_4$ (g)	Final activity (UEA)	Isolation yield (%)
1.	*Candida rugosa*, 24 hr	20	3 820	14	2 368	62.0
2.	*Candida rugosa*, 28 hr	800	289 600	560	85 721	29.6
3.	*Yarrowia lipolytica*, 24 hr	20	2 320	14	2 204	95.0
4.	*Yarrowia lipolytica*, 28 hr	800	85 200	560	80 940	95.0

Table 3. The final activities and isolation yields determined for the crude lipases separated from the cultivation media of the yeasts strains *Candida rugosa* DSM 70761 and *Yarrowia (Candida lypolitica)* ATCC 8661

At a first step the preliminary transesterification results obtained by thin layer chromatography demonstrated both lipases have high enough catalysis activities.

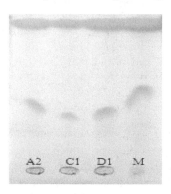

Legend:
A2-lipase from *Candida rugosa DSM 70761*/M2; C1-lipase from *Candida rugosa DSM 70761*/M3; D1-lipase from *Yarrowia lipolytica*/M2; M - Control, ester of oleic acid

Fig. 4. Thin layer chromatography of the products obtained by the transesterification

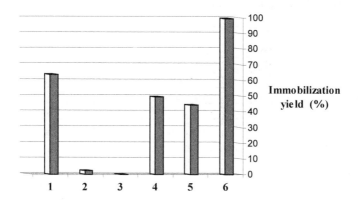

Fig. 5. Immobilization efficiency of the tested lipases (Tcacenco *et all*, 2010)

The experimental results are presented in the Figure 5, obtained with the described immobilization techniques for both crude lipases.
The immobilization techniques, characterized in the following table, were performed in comparison with the immobilization of a lipase from the fungus *Aspergillus niger*.

No	Lipase source	Immobilization technique
1	*Candida rugosa,* DSM 70761	Chitosan adsorption and cross-linking with glutaraldehyde
2	*Aspergillus niger* (Fluka)	Chitosan adsorption and cross-linking with carbodiimide
3	*Aspergillus niger* lyophilized lipase (Fluka)	Chitosan adsorption, cross-linking with glutaraldehyde and granulation with sodium borohydride
4	*Candida rugosa,* DSM 70761	Adsorption on Celite 545
5	*Candida rugosa,* DSM 70761	Adsorption on Silicagel G
6	*Yarrowia lipolytica* ATCC 8661	Adsorption on Celite 545

Table 4. Applied immobilization techniques

The experimental study regarding the immobilization of lipases gave interesting results: high yield of 99% obtained for the immobilization of *Yarrowia lipolytica* lipase by adsorption on Celite support, good yields of 63.26% for the immobilization of *Candida rugosa* lipase by adsorption on chitosan cross linked with glutaraldehyde and respectively 44 - 49% for the same lipase immobilized by adsorption on Celite or Silicagel. On the contrary the immobilization of *Aspergillus niger* lipase gave unsatisfactory results.
In order to improve the immobilization yield of the lipase from the yeast *Candida rugosa* DSM 70761 on Celite support a supplementary treatment with acetone as organic solvent

was done, the obtained results in comparison with the control procedure (without acetone adding) are presented in the following table.

No.	Support	Lipase (mL)	Initial activity		Final activity		Immobilization yield (%)
			UEA/mL	Total activity	UEA/mL	Total activity	
1.	Celite 545	800	362.0	289 600	840.2	142 841.7	49.32
2.	Celite 545 + acetone	1150	182.6	210 000	2537.2	204 246.2	97.26

Table 5. The immobilization yields of the lipase from the yeast *Candida rugosa* DSM 70761 on Celite support with / without acetone treatment

The acetone treatment had as consequence a big improvement of the immobilization yield on Celite from 49% to 97% in case of the lipase from *Candida rugosa* DSM 70761. It seems that the system hydration degree highly increases due to the support treatment with organic solvent, which determines a better adsorption of the enzyme. The improved procedure to get the immobilized biocatalyst was further applied in the research regarding the immobilized enzyme characteristics: static activity, operational activity, and transesterification performance.

The immobilized lipases from both yeasts *Yarrowia lipolytica* and *Candida rugosa* prepared by Celite adsorption were preserved in a freezer at -18°C and tested for static stability at different time duration. Results are presented in the Figure 6.

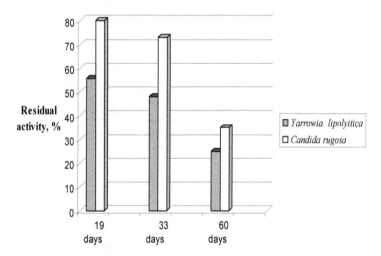

Fig. 6. Static stability determination by yield of the residual activity for the Celite adsorption immobilized lipases from *Yarrowia lipolytica* and *Candida rugosa*

The results demonstrated a higher static time stability for the Celite adsorption immobilized lipase from the yeast *Candida rugosa* DSM 70761, with 73% residual activity after more than 1 month, by comparison with only 48% residual activity for the immobilized lipase from the yeast *Yarrowia lipolytica* ATCC 8661.

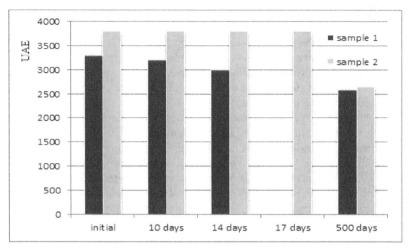

Fig. 7. The effect of the acetone treatment associated to Celite immobilization on the static activity of the immobilized lipase from *Candida rugosa* DSM 70761 (sample 1-no acetone treatment; sample 2-with acetone treatment)

The static stability for the enzyme from *Candida rugosa* DSM 70761 immobilized on Celite with / without acetone treatment was determined for a long period of time, the results being presented in the Figure 7. The biocatalyst obtained with the lipase from the above mentioned yeast immobilized on Celite 545 by physical adsorption with or without organic medium treatment presented a high static stability, when preserved in freezer. The residual activity was as high as 82% after 1 year and half, and after the first 2 weeks the residual activity was practically unchanged in both cases. These findings were considered as a selection criterion between the lipases from the two studied yeasts, so the lipase produced by *Candida rugosa* DSM 70761 with a better static stability was further used to continue the research. Firstly the biocatalyst prepared by the described procedure with the lipase from the yeast *Candida rugosa* DSM 70761 was tested for its operational activity. The test consisted of using the same biocatalyst quantity in several reaction cycles and measuring the enzymatic activity at the beginning and after each reaction phase. The results are presented in the following table.

No. of cycles	Initial activity (UEA/g)	Final activity (UEA/g)	Activity loss (%)
1.	705.00	606.72	13.34
2.	606.72	549.75	9.40
3.	549.75	481.36	12.44
4.	431.36	413.97	14.00

Table 6. Evolution of the operational activity of the lipase from *Candida rugosa* DSM 70761 immobilized on Celite 545

The results from the table indicate a biocatalyst half time of 5-6 reaction cycles, because after 4 cycles the residual activity was 58.71%.

3.3 Lipase transesterification
3.3.1 Materials and methods

The experimental study was done with rapeseed oil of Romanian origin or soya oil and by using the lipase from the yeast *Candida rugosa* DSM 70761 obtained in aerobic bio processing, isolated from the cultivation medium and immobilized by adsorption on Merck Celite support (lipase activity of 4701 UEA / g support).

The transesterification was done in two variants: (a) in anhydrous medium without organic solvents adding; (b) in hexane (Biosolve).

The experimental working procedure was: the transesterification reaction was performed in Erlenmayer flasks of 100 mL, containing the tested vegetable oil in a concentration to determine a final triglycerides content of 0.08 mol / L and methanol (this last reagent in molar ratios between 3:1 and 8:1 with the triglycerides substrate). The immobilized enzyme was added in a chosen concentration after a period of 30 minutes at 37°C. The reaction was done with continuous mixing of 250 rpm. Each 4 hours' sample from the liquid was analysed by thin layer chromatography and gas chromatography to determine the reaction advancement.

a. Thin layer chromatography was done by using the Silicagel G on Al support as stationary phase and the migration solvent was a mixture of petroleum ether: ethylic ether: acetic acid = 80:30:1; the spots were put into evidence in a iodine vapour atmosphere.

b. The fatty acids content in methylic esters was analysed by gas chromatography (GC) using a capillary column with a stationary phase composed from 5% phenyl – 95% methylpolysilane.

The apparatus was a gas chromatograph 6890N – AGILENT with FID detector and autosampler 7683B; column HP 5, L=30m; φ=0.32mm.

Reagents: N-hexane; the methylic esters of several fatty acids mostly presented in the vegetable oils (rapeseed oil or soya oil).

Working conditions:

- column temperature: initial temperature of 160°C, during 2min.; final temperature of 240°C, during 5min.; heating rate of 5°/min.
- injection temperature of 280°C.
- detector temperature of 300 °C.
- nitrogen flowrate of 2.0 mL / min.
- hydrogen flowrate of 40 mL / min.
- air flowrate of 370mL / min.
- nitrogen flowrate (make-up) of 25 mL / min.
- sample volume of 1µl.
- analysis duration of 23 min.

The evaluation is done by the determination of the content in palmitic, oleic, arachidonic and erucic acids. The external standard method is applied.

Three experimental models were studied:

a. Batch enzymatic transesterification with methanol and without organic solvent, characterised by : vegetable oil concentration of 0.09 M; methanol concentration of 0.54 M (ratio of 8:1 methanol: triglycerides substrate); biocatalyst concentration of 5000 UEA / 100 mL reaction medium; reaction temperature of 37°C; mixing of 250 rpm; reaction total duration of 24 hr.

b.   "Semi-batch" enzymatic transesterification with methanol and without organic solvent, characterised by the same reaction conditions, except the fact that the methanol is added two times, each addition realizing a ratio between alcohol and the triglycerides substrate of 4:1.

c.   Batch enzymatic transformation in hexane characterized by: vegetable oil concentration 0.09M; methanol concentration 0.09 M; biocatalyst concentration 5000 UEA / 100 mL reaction medium; reaction temperature of 37⁰C; mixing of 250 rpm; reaction total duration of 24 hr.

3.3.2 Results and discussion

The most important transesterification results are presented in the Figure 8.

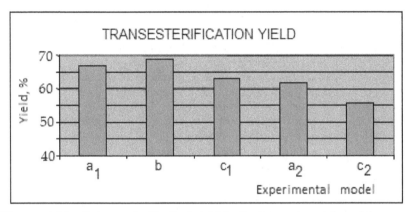

Fig. 8. The vegetable oils transesterification yield for the 3 experimental models: a_1-batch model without solvent, with soya oil; a_2-batch model without solvent, with rapeseed oil; b-semibatch model without solvent, with soya oil; c_1-batch model in hexane with soya oil; c_2-batch model in hexane with rapeseed oil

The transesterification was realised with the following biotransformation yields: 56-67 % for the experimental variant a, 69% for the variant b, and 56-63% for the variant c; these results can be improved by adequate optimization procedures to be applied for each technological phase, comprising enzyme obtaining in aerobic bioprocess, lipase immobilization and transesterification performing.

4. Conclusions

1.   The main objectives of the research to replace the actual chemical transesterification with the enzymatic process are: (a) the preparation of cheap and stable immobilized lipases; (b) the realization of biotransformation systems characterized by the biocatalyst long use in many reaction cycles. One of the raisons to choose between extracellular or intracellular lipases is the immobilization of extracellular enzymes by physical adsorption, a low price technology, but imposing to improve the shorter duration. Moreover these lipases are normally biosynthesized by bacteria or yeasts, easier to cultivate in aerobic bioprocess than the intracellular lipases producing' fungi.

2. The lipases with advanced specificity are not useful in the transesterification to produce biodiesel; the most recommended are the lipases with reduced region specificity, but more developed specificity for the substrate.
3. The molar ratio of the substrates used in the biotransformation of vegetable oils to biodiesel must be determined for each studied system: alcohol – oil – lipase.
4. The rapeseed oil is of interest as raw material in the transesterification, as it is largely produced by the European agriculture and also in Romania, and at the same time it is used in the alkaline catalysed transformation. But in USA the soya oil is in charge.
5. The aerobic bio processing of several bacteria and yeasts from Romanian research collections or from international collections demonstrated that two yeasts, *Candida rugosa* DSM 70761 and *Yarrowia lipolytica* ATCC 8661 produced lipases characterized by high activity in simple and short duration cultivation. The media composition and the cultivation parameters were optimized for both yeasts' lipases formation.
6. The immobilisation techniques by physical adsorption were studied for the lipases from the above mentioned yeasts. First of all the extracellular lipases from the yeasts *Candida rugosa* DSM 70761 and *Yarrowia lipolytica* ATCC 8661 can be easily separated in the liquid fraction by centrifugation and further on the crude enzymes can be obtained by ammonium sulphate precipitation. The experimental study regarding the immobilization of lipases gave interesting results: high yield of 99% obtained for the immobilization of *Yarrowia lipolytica* lipase by adsorption on Celite support, good yields of 63.26% for the immobilization of *Candida rugosa* lipase by adsorption on chitosan cross linked with glutaraldehyde and respectively 44 - 49% for the same lipase immobilized by adsorption on Celite or Silicagel.
7. As the lipase from the yeast *Candida rugosa* DSM 70761 was immobilized on Celite 545 support with yields of 49 – 63%, and higher yields are obtained for the immobilization of the lipase from *Yarrowia lipolytica* ATCC 8661, and the immobilization procedure is easy and low price, the laboratory experimental model was developed on this support.
8. In order to improve the immobilization of the lipase of *Candida rugosa* DSM 70761, a treatment with acetone as organic solvent was introduced and this operation had as consequence a big increase of the immobilization yield on Celite from 49% to 97%.
9. A higher static stability was determined for the Celite adsorption immobilized lipase from the yeast *Candida rugosa* DSM 70761, with 73% residual activity after more than 1 month, by comparison with only 48% residual activity for the immobilized lipase from the yeast *Yarrowia lipolytica* ATCC 8661. The residual activity was as high as 82% after 1 year and half, and after the first 2 weeks the residual activity was practically unchanged for the first biocatalyst. These findings were considered as a selection criterion between the lipases from the two studied yeasts, so the lipase produced by *Candida rugosa* DSM 70761 with a better static stability was further used to continue the research.
10. This biocatalyst operational stability was also tested and the immobilized enzyme half time was of about 5-6 reaction cycles, as after 4 reaction cycles the residual activity was still 58.7%.
11. Three experimental models were considered to perform the transesterification: (a) batch enzymatic transesterification with methanol and without organic solvent; (b) "semi-batch" enzymatic transesterification with methanol and without organic solvent; (c) batch enzymatic transformation in hexane. The reaction yields were good enough for all the tested experimental models and for both -soya and rapeseed oils, the results variation being in the range of 56 - 69%. They can be improved by adequate

optimization procedures to be applied for each technological phase, comprising enzyme obtaining in aerobic bioprocess, lipase immobilization and transesterification. Further research work is to be developed in two directions: (1) the use of the glycerol formed as by product in the transesterification process, especially as C source in several other bioprocesses; (2) as beside this product there are several others, the most important future research direction will be the technical application of the bio refinery concept realised for the vegetable oils extracted from many plants specific to each geographical area. A possible future bio refinery will integrate physical, chemical, and biological procedures for the biodiesel preparation, conversion of solid residue with high carbohydrates or protein content; glycerol use, the whole application being characterised by both high economic efficiency and reduction of solid or liquid residues.

5. Acknowledgment

This research was financially supported by the Romanian National Agency for Research under the Second National Program, Project 61-032/2007.

6. References

Antczak M. S., Kubiak A., Antczak T, & Bielecki S. (2009), Enzymatic biodiesel synthesis – Key factors affecting efficiency of the process, *Renewable Energy*, Vol. 34, 1No. 4, pp.1185–1194, ISSN 0960-1481

Bajaj A., L. Purva, Jha P. N., & Mehrotra R., (2010), Biodiesel production through lipase catalyzed transesterification: An overview, *Journal of Molecular Catalysis B: Enzymatic,* Vol. 62, No. 1, pp. 9–14, ISSN 1381-1169

Bisen P. S., Sanodiya B. S., Thakur G. S., Baghel R. K., & Prasad G. B. K. S., (2010), Biodiesel production with special emphasis on lipase-catalyzed transesterification, *Biotechnological Letters*, Vol. 32, No. 10, pp. 1019-1030, ISSN 0141-5492

Chirvase A. A., Ungureanu C., Tcacenco L., Radu N., (2010), Determination of Yeast Strains Characteristics as Lipase Providers for Enzymatic Transesterification to Biodiesel *Revista de Chimie*, Vol. 61, No.9, pp. 866-868, ISSN 0034-7752

Fjerbaek L., Christensen K.V., & Norddahl B., (2009), A Review of the Current State of Biodiesel Production Using Enzymatic Transesterification, *Biotechnology and Bioengineering*, Vol. 102, No. 5, April 1, pp. 1298-1315, ISSN 1097-0290

Ghaly A. E., Dave D., Brooks, & Budge S., (2010), Production of Biodiesel by Enzymatic Transesterification: Review, *American Journal of Biochemistry and Biotechnology*, Vol. 6, No 2, pp. 54-76, ISSN 1553-3468

Moore A., (2008a), Biofuels are dead: long live biofuels (?) – Part one, *New Biotechnology*, Vol. 25, No. 1, pp. 6-13, ISSN 1871-6784

Moore A., (2008b), Biofuels are dead: long live biofuels (?) – Part two, *New Biotechnology*, Vol. 25, No. 2/3, pp. 96-101, ISSN 1871-6784

Tcacenco L., Chirvase A. A.,Berteanu E., (2010), The preparation and immobilization of some yeast lipases for rapeseed oil transesterification to biodiesel, *Romanian Biotechnological Letters*, Vol. 15, No. 5, pp. 5631-5639, ISSN 1224-5984

Uthoff S., Broker D., Steinbuchel A., (2009), Current state and perspectives of producing biodiesel-like compounds by biotechnology, *Microbial Biotechnology*, Vol. 2, No. 5, pp. 551-565, ISSN 1751-7915

Adsorption in Biodiesel Refining - A Review

Carlos Vera, Mariana Busto, Juan Yori, Gerardo Torres,
Debora Manuale, Sergio Canavese and Jorge Sepúlveda
INCAPE (FIQ, Universidad Nacional del Litoral-CONICET),
Argentina

1. Introduction

Biodiesel is a petrodiesel substitute composed of a mixture of fatty acid methyl esters obtained by the transesterification of plant oils or animal fats with short chain alcohols such as methanol or ethanol. Despite its natural origin biodiesel is technically fully compatible with petroleum diesel, requiring virtually no changes in the fuel distribution system or the Diesel motor. Its production and use have increased significantly in many countries and are in nascent status in many others. Other advantages of biodiesel compared to petrodiesel are reduction of most exhaust emissions, biodegradability, higher flash point, inherent lubricity and domestic origin (Chang et al., 1996; Romig & Spataru, 1996; Wang et al., 2000).

Literature on the refining of biodiesel is abundant but concentrates almost exclusively on the transesterification steps for transforming fats and oils into esters of short alcohols and fatty acids. In this sense in the last years the most important advances in the reaction technology have been the development of continuous heterogeneous transesterification reactors (Bournay et al., 2005; Portilho et al., 2008) and the design of new robust non-catalytic processes for multifeedstock operation (Saka & Kusdiana, 2001; Saka & Minami, 2009).

In the case of the refining operations downstream and upstream the transesterification reactors the biodiesel literature is however scarce. Two are the reasons for this: (i) Feedstock pretreatment in the case of biodiesel is a mature technology developed decades ago for the production of edible oil. (ii) After natural triglycerides are converted into fatty acid methyl esters, the product mixture needs little chemical adjustment since many properties of these esters are ideal for the functioning of Diesel motors.

Some reports on post-reactor biodiesel refining have dealt with classical and simple techniques of purification, e.g. water washing (Karaosmanoglu et al., 1996). Others have indicated that adsorption technologies are particularly suited for the refining of biodiesel (Yori et al. 2007; Mazzieri et al., 2008; Manuale et al. 2011). In order to elucidate the role of adsorption processes in the refining of biodiesel, this review studies some theoretical and practical aspects related to the functioning, design and operation of adsorbers and their application to the purification of biodiesel product and feedstocks.

2. The needs for refining of petrodiesel and biodiesel fuels

The objectives of Diesel fuel refining operations are aimed at improving the fuel combustion performance, maximizing the power delivered to the motor, increasing the engine life and reducing the emission of noxious compounds. The relevant properties involved are cetane index, heat content, lubricity, viscosity, cold flow properties, oxidation stability and amount and kind of tailpipe emissions. Some properties are superior for biodiesel in comparison to petrodiesel and need no adjustment. This is the case of lubricity (93% film for biodiesel and 32% for petrodiesel), cetane index (45 for petrodiesel and 56 for biodiesel) and tailpipe emissions (Chang et al., 1996; Romig & Spataru, 1996). Other properties of biodiesel needing adjustment will be discussed in the next paragraphs.

Viscosity. Viscosity affects injector lubrication and atomization. Natural oils and fats (triglycerides) have excessive viscosity and cannot be easily injected; this is the main reason why they must be transformed into methyl esters. Even after transesterification the viscosity of biodiesel is higher than that of petrodiesel (5 cSt at 40 °C compared to 3 cSt), though it is considered enough low in international norms. A few reports have however indicated that additivations and chemical transformations could advantageously alter biodiesel viscosity. Noureddini et al. (1998) found that addition of GTBE in amounts as big as 22% could not only lower the pour and cloud points of biodiesel but also its viscosity by 8%. Yori et al. (2006) studied the acid-catalyzed isomerization of methyl soyate and found that isomerization decreased pour and cloud points but adversely increased the viscosity.

Oxidation stability. The stability of a diesel fuel is related to the occurrence of undesired reactions during storage. In the case of petrodiesel routine oxidation tests as performed by ASTM D2274 detect the formation of minor amounts of insolubles that are due to the precipitation of polar compounds, mainly polycyclic acids, after their reaction with iron particles or oxygen (Díaz & Miller, 1990). In the case of biodiesel the problem is worse because unsaturated fatty acid chains are main components of the fuel and they are active in oxidizing reactions. In contact with oxygen, peroxides are formed that promote the formation of organic acids, and then of polymers (gums) that plug fuel lines and filters. Oxidative degradation during storage can also compromise fuel quality with respect to effects on kinematic viscosity, acid value, cetane number, total ester content, and formation of hydroperoxides, soluble polymers, and other secondary products (Du Plessis et al., 1985; Bondioli et al., 2002; Thompson et al., 1998). The increased acidity and peroxide values as a result of oxidation reactions can also cause the corrosion of fuel system components, the hardening of rubber components and the erosion of moving parts (Tang et al., 2008). By now the only method for increasing the biodiesel resistance to oxidation is to add synthetic or natural oxidation inhibitors such as tocopherols and hydroquinones. Other alternative way is the hydrogenation of the unsaturated chains. Compared to untreated soybean oil methyl esters, partially hydrogenated products have shown superior oxidative stability and similar specific gravity, but inferior low-temperature performance, kinematic viscosity and lubricity (Moser et al., 2007). In order to raise the saturated fraction of biodiesel other efforts have been carried out by distillation and crystallization (Falk & Meyer-Pittroff, 2004) and it is conceivable that the same could be done by adsorption over suitable materials.

Storage stability. Also related to the stability of biodiesel, some other minor components of biodiesel, the monoglycerides (MGs) and diglycerides (DGs) can form crystals during storage at low temperatures and precipitate. These crystals not only can clog fuel lines and fuel filters but due to their amphiphilic nature, their absence in the solution causes the precipitation of other unstable solvatable impurities such as glycerol.

Acidity. Acidity in petrodiesel is mainly related to the presence of napthenic acids in the crudes. Acidity of biodiesel depends on a wider variety of factors and is influenced by the type of feedstock used and on its degree of refinement. Acidity can also be generated during the production process, e.g. by mineral acids introduced as catalysts or by free fatty acids resulting from hydrolysis of soaps and esters. Biodiesel acidity also reflects the degree of fuel ageing during storage, as it gradually increases due to hydrolytic cleavage of ester bonds. High fuel acidity of biodiesel has been discussed in the context of corrosion and the formation of deposits within the engine, particularly in fuel injectors, by catalyzing polymerization in hot recycling fuel loops (Refaat, 2009). However the main problem associated with acids is the formation of soaps as it will be discussed later.

Carbonization properties. Formation of carbon deposits in the injectors of a Diesel engine is undesired; the tendency of a fuel to form these deposits being measured by the Conradson Carbon Residue (CCR) test (ASTM D189). In the case of petrodiesel CCR is related to the presence of aromatic and polyaromatic compounds, and is favorably reduced by hydrotreatment. In the case of biodiesel deposits formed in the injectors are related to polymerization of glycerol and glycerides. These polymers undergo further decomposition to carbon deposits and tarnishes over injectors and cylinders. In this sense the needed biodiesel refining step is the removal of free and bound glycerol to minimum values. ASTM D6751 constrains the iodine number of biodiesel to less than 112 on the same basis because olefinic chains are also reactive for polymerization. However this is not an issue in european norms (EN 14214).

Cold flow properties. In raw biodiesel the presence of wax-like, long acyl chains, poses the problem of crystallization when temperature is too low. Crystal nucleation is enhanced by the presence of MGs and DGs, mainly affecting the cloud point (van Gerpen et al., 1996). A first solution is to eliminate glycerides to negligible values. Other solutions for waxy FAMEs are not without drawbacks: (i) Catalytic dewaxing (Yori et al., 2006) decreases the cetane number and increases the viscosity. (ii) Winterization for removing the waxy saturated fraction also removes the fraction with higher cetane and oxidizing stability. (iii) Commercial pour point depressants are reported to reduce the pour point of biodiesel but usually do not reduce its cloud point nor improve its filterability at low temperatures (Dunn et al., 1996). Fortunately, biodiesel-petrodiesel blends have cloud and pour points closer to those of petrodiesel. The saturated portion eliminated by winterization can also be used as a "summer fuel" if massive storage is available.

Refinery operation issues. Some specifications for feedstocks and intermediate streams in refineries are related to the correct functioning of process units. Sulfur reduction in the case of petroleum fuels is necessary not only to improve the quality of the final product but also to prevent the poisoning of catalysts in some hydroprocessing units (Ito and van Veen, 2006). In the case of biodiesel many undesired components are responsible for the malfunctioning of reactors and phase separators:

- Phosphorous, calcium, and magnesium are minor components typically associated with phospholipids and gums that act as emulsifiers or cause sediments, lowering yields during the transesterification process. Phosphorus typically leads to an increased difficulty in the separation of the biodiesel and glycerol phases (Anderson et al., 2003).

Component	Crown Iron Works (USA)	Lurgi GmbH (Germany)
Moisture and volatiles	0.05% max	0.1% max
Acidity	0.5% max	0.1% max
Phosphorus total	20 ppm max	10 ppm max
Soap	50 ppm max	n.a.
Unsaponifiables	1% max	0.8% max

Table 1. Quality requirements for the feedstock of two alkali homogeneous catalyst biodiesel production technologies (Anderson et al., 2003; Lurgi, 2011).

- FFAs and soaps. In the case of the alkali-catalyzed process, the dominant biodiesel technology, the presence of free fatty acids (FFAs) leads to the use of an increased amount of catalyst and other chemicals. It also increases the concentration of salt and water in the crude glycerol phase. Aside from the increased cost of chemicals, the presence of FFA causes a larger potential for soap formation and all the production issues associated with soap, including more difficult phase separations and more frequent cleaning of process vessels. Although FFA can be reacted in an acid-catalyzed reaction with methanol to form methyl esters, the amount of acid required is much higher than the amount of catalyst used in the transesterification of neutral oil. The reaction also does not go as far to completion as transesterification, which may lead to the resulting biodiesel product to be out of specification on FFA. Acid catalysis of FFA to methyl esters also results in higher salt and water formation. For all these reasons feedstock specifications for FFA have low limits.
- Unsaponifiable matter (UM) consists of plant sterols, tocopherols and hydrocarbons, with very small quantities of pigments and minerals. UM is limited in the feedstock of biodiesel processes mainly on the basis of its foaming properties that make separations difficult (see Table 1). The unsaponifiable matter is not affected by ester preparation, so it is likely to be present in similar amounts in biodiesel to its level in the crude feedstock. UM has no harmful effects except possibly for a change in the crystallization onset temperature (van Gerpen et al., 1996). For this reason UM is not limited in biodiesel norms. Some unsaponifiable compounds, such as the phytosterols, have antioxidizing capacities and they are useful for prolonging the storage life of biodiesel (Rabiei et al., 2007). A possible challenge for adsorption operations in this case could be the selective removal of impurities while not affecting these antioxidizing compounds.
- Water. Alkaline catalysts (NaOH, KOH, MeONa) react with water and oil to produce soaps. Acid catalysts (e.g. H_2SO_4) when hydrated reduce their effective acid strength and their catalytic activity. Water thus leads to deactivation and higher catalyst usage.

3. Refining of biodiesel feedstocks

Depending on their degree of refining, biodiesel feedstocks might need some or all the refining steps common to the refining of edible oils: (i) Degumming, that is necessary if large amounts of phosphatides are present in the feedstock, phosphoric acid and steam being used to swell the gums for further removal. (ii) Deodorization, that is used with feedstocks up to 30% FFA. It is basically a vacuum distillation at 240-270 °C and 2-5 mmHg, that removes aldehydes, ketones and smelly products, pesticides, fungicides, herbicides, etc. It also lightens up the product by destroying carotenoids. (iii) FFA reduction by many means, steam stripping, caustic stripping, solvent extraction, glycerolysis, acid esterification, etc. (iv) Bleaching, that is normally used to remove remaining impurities such as pigments, soaps, insolubles, peroxides, phospholipids and metals.

It must be noted however that biodiesel and edible oil have different quality specifications. This is especially true for color and odor, that are not an indication of technical quality of biodiesel. ASTM quality biodiesel can range from clear to black and have an unpleasant smell. This is more a consumer issue because it raises uncertainties. Color removal may need carbon filtration and bleaching while odor removal may need deodorization.

The degree of FFA reduction in biodiesel feedstocks needs special attention because it has a high dependence on the technology of biodiesel production used: (i) In the case of the non-catalytic method that uses supercritical methanol at high temperatures and pressures, FFA content is not an issue, because triglycerides and FFAs react to form methyl esters with similar rates (Warabi et al., 2004). (ii) In the acid-catalized method feedstocks with up to 20% acidity can be completely reacted by acid catalysis with mineral acids though the kinetics are much slower (Lotero et al., 2005; Freedman, 1986). (iii) The alkali-catalized method (dissolved NaOH, KOH, MeONa, etc. catalysts) tolerates only 0.5% FFA in the feedstock (Table 1). However some producers accept feedstocks of up to 4% FFA and then use caustic stripping by the same catalyst in the reactor or caustic washing before the reactor to eliminate them from the reaction medium. The soap that goes into the glycerol phase or the wash water is hydrolized and reacted to biodiesel by acid-catalyzed transesterification in a separate reactor. (iv) In the acid-base method feedstocks with up to 20% acidity are first esterified in acid catalysis and then the reaction is continued with alkaline catalysis.

The use of adsorbents for the pretreatment of biodiesel raw materials is related to known techniques for edible oil refining. After pressing of oil seeds, and after degumming and caustic refining of the virgin oil, a step of bleaching is commonplace in order to improve the colour by adsorbing chlorophylls, carotenoids and other pigments, and the removal of other undesired components such as metals and free fatty acids, that contribute to the unstability of the oil under oxidizing conditions. Bleaching of oils can be done with natural clays such as bentonite, smectite, montmorillonite, etc., or activated clays produced by acid treatment (Foletto et al., 2011). Clays are mainly used for removing high molecular weight organic compounds but their affinity for polar compounds and metals is low. In this sense most part of the metals is eliminated during the caustic refining of edible oils and in the subsequent water washing steps, while bleaching with clays does not practically modify the metal content (van Dalen & van Putte, 1992; Farhan et al., 1988).

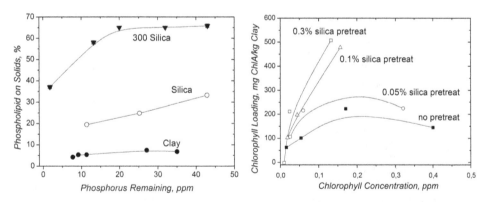

Fig. 1. Isotherms of adsorption of phosphatides on silica (left). Adsorption of chlorophyll on clay as a function of the silica adsorption pretreatment (right). Welsh et al. (1990).

Another adsorbent commonly used is silica, either alone or together with clays, though it is now accepted that best treatments should include some portions of both, since silicas adsorb preferably polar compounds and clays are more suitable for organic compounds. Treatment with silica has become incresingly widespread and silicas for edible oil refining have become highly tailored for this application, thus leading to the coining of the term "silica refining" (Welsh et al., 1990). The conditions for optimal silica refining can be summarized as follows: (i) Oil temperature is raised to 70-90 °C. (ii) Silica is added at atmospheric pressure to the vessel contaning the oil. (iii) The moisture content of the oil is reduced to 0.2-0.5% by evaporation, preferably in a vacuum. (iv) The contact time between the silica and the oil should be 10-15 min. (v) The moisture in the oil plays an important role in the mechanism responsible for transporting polar compounds from the oil to the silica, where they are trapped. (vi) After the removal of the polar contaminants the oil should be further dried if clays are to be used in the bleacher. During the vacuum drying process water is removed from the silica and the weight is reduced to even 40% of its original value; the solid reduces also in size and so does the load on the filters downstream the bleachers, which can then be operated at higher filtration flowrates and longer filtration cycles.

As silicas are far more efficient adsorbents for polar contaminants, if colour is not an issue (like in the case of biodiesel fuel) they can easily replace bleaching clays. If colour reduction is necessary then clays can be used in a second step after silicas have removed the polar contaminants. This reduces the amount of adsorbent used and enhances the quantity of oil produced because a lower quantity of filter cake is produced and oil losses are reduced. In this sense, a common industry perception is that 20-25% of oil is present in the filter cake but as oxidized and polymerized oil are not extracted in the extraction tests, the typical oil content in the cake can be as high as 40%.

The claimed advantages of silica (Grace, 2011) for refining biodiesel feedstocks are: (i) Lower costs of residue treatment by means of the reduction of effluents. (ii) Lower costs by elimination of washing steps. (iii) Lower product losses. (iv) Higher yield of the biodiesel fuel precursor. (v) Lower demand of catalyst in the transesterification reactor due to a lower FFA content. (vi) Lower consumption of acid for neutralization of the catalyst. (vii) Higher yield of biodiesel due to an enhanced separation of the glycerol and biodiesel phases (absence of soaps and glycerides). (viii) Purer glycerol due to a low content of impurities. (ix) Lower costs of production of biodiesel. (x) Quality improvement due to an enhanced stability (absence of metals and FFA).

One additional benefit of silica addition in the case of the caustic refining for oil treatment (e.g. for biodiesel alkaline processes of low FFA tolerance) is that water-wash centrifuges can be eliminated because silicas efficiently remove residual metals, phospholipids and soaps. These must be otherwise washed away to prevent them reaching the bleaching units.

4. Refining of crude biodiesel

After transesterification is completed, many contaminants can still be present in the biodiesel product depending on the technology of transesterification used (Table 2). Removal of these impurities will be treated separately in the next subsections.

4.1 Glycerides

Removal of glycerides from biodiesel is an important step of the process because key aspects of the quality of the fuel strongly depend on the content of bound glycerol. The ASTM D6751 and EN 14214 standards establish a maximum amount of 0.24-0.25% bound glycerol. Main problem with these compounds is that when heated they tend to polymerize forming deposits. They also increase the cloud point of biodiesel and they complicate the operation of liquid-liquid phase splitting units due to their amphiphilic nature.

Impurity	Alkali-catalyzed	Acid-catalyzed	Supercritical
Soaps	Yes. By neutralization of FFA with catalyst	After neutralization of the catalyst	If feedstock treatment was uneffective.
Metals, P	If feedstock treatment was uneffective.	If feedstock treatment was uneffective.	If feedstock treatment was uneffective.
FFAs	No	Yes. Due to incomplete esterification.	Yes. Due to hydrolysis of the feedtock.
Monoglycerides	Yes. Product of transesterification.	Yes. Product of transesterification.	Yes. Product of transesterification.
Diglycerides	Yes. Product of transesterification.	Yes. Product of transesterification.	Yes. Product of transesterification.
Triglycerides	Yes. Due to incomplete conversion.	Yes. Due to incomplete conversion.	Yes. Due to incomplete conversion.
Glycerol	Yes. Product of transesterification.	Yes. Product of transesterification.	Yes. Product of transesterification.

Table 2. Contaminants in biodiesel product depending on transesterification technology.

In the specific case of monoglycerides (MG), diglycerides (DG) and triglycerides (TG), they are the raw materials and the intermediates of the transesterification reaction. This is an equilibrium reaction with an equilibrium constant close to unity (D'Ippolito et al., 2007), that dictates that a methanol excess must be used to shift the equilibrium to the right and to decrease the concentration of triglycerides and intermediates in the final product mixture. Noureddini & Zhu (1997) and Darnoko & Cheryan (2000) studied the kinetics of transesterification of oil and they reported that the conversion value for the 1-step reaction of transesterification of soy oil with methanol in a stirred tank reactor, using a methanol-to-oil ratio of 6 was 80-87% at 1 h of time of reaction. Busto et al. (2006) indicated that in supercritical tubular reactors a methanol-to-oil ratio of 6 yields an equilibrium value of 94-95% at high Péclet numbers. In the case of processes with two reaction steps, after the final step of glycerol removal, the amount of TG, MG and DG is sufficiently low to almost comply with the ASTM D6751 limits. It can be however deduced that this final content of bound glycerol is a function of the methanol-to-oil ratio used in the reaction and the number of reaction steps. For the alkali catalized process with two reaction steps this methanol-to-oil ratio is 6. In the case of the supercritical method with one reaction step (Goto et al., 2004) the adequate methanol-to-oil molar ratio is reported to be 42. The final adjustment of the glycerides content is made in the standard industrial practice by water washing. Some authors however propose separating the glyceride fraction (Goto et al., 2004; D'Ippolito et al., 2007) and recycling it to the reactor.

One interesting issue is that of the relative concentration of MG, DG and TG in the final product. According to data of Noureddini and Zhu (1997) the equilibrium constants for the partial transesterification (producing 1 mol of FAME) of triglycerides, diglycerides and monoglycerides are $K_1=0.45$, $K_2=0.18$, $K_3=34.6$. TGs would therefore be thermodynamically more stable. It is however found in practice, probably because of kinetic limitations, that MGs and DGs are main impurities (He et al., 2007). This points to the adequacy of adsorption treatments since MGs are efficiently removed by adsorption over silica, even in the presence of water and soaps (Mazzieri et al., 2008).

Some points seem clear: (i) The final bound glycerol content is a function of the methanol-to-oil ratio. (ii) An adequate separation/recycling or removal/disposal of glycerides could reduce the complexity of the process by reducing the methanol-to-oil ratio and the amount of recycled methanol. (iii) MG and DG should be the focus for reducing bound glycerol.

Steps in the direction of (ii) have been hinted by D'Ippolito et al. (2007) and Manuale et al. (2011) for the supercritical method. The first proposed using 2 reaction steps with a low methanol-to-oil ratio (6-10), retaining glycerol and glycerides in packed bed adsorbers and recycling them to the reactor. The second indicated that the combination of 1-step reaction, a methanol-to-oil ratio of 15-20 and silica refining could produce EN14214 grade biodiesel.

4.2 Glycerol

Liquid-liquid equilibrium studies of biodiesel-methanol-glycerol mixtures have been undertaken in the past by Kimmel (2004), Negi et al. (2006) and Zhou & Boocook (2006). They determined that the equilibrium glycerol content in biodiesel depends strongly on the residual content of methanol acting as a cosolvent. When methanol is completely removed the free glycerol content depends only on the temperature, being approximately 0.2% at 25 °C and increasing linearly with temperature (Kimmel, 2004). Even if methanol is not present hydrophilic glycerol can be solubilized in the oil phase by amphiphilic MG and DG. These glycerides can separate from the oil during storage and precipitate as a result of temperature changes or long residence times. Glycerol then precipitates as a consequence of the reduced solubility, leading to the formation of deposits. Soluble glycerol is also a problem because glycerol polymerizes on hot surfaces (cylinders, injectors) with formation of deposits or "tarnishes". For all these reason glycerol should be thoroughly removed.

Glycerol removal by adsorption was early performed by Griffin and Dranoff (1963) using sulfonic resin beads. Glycerol adsorption over polar surfaces is favored if dissolved in organic media that have little affinity for the adsorbent. Nijhuis et al. (2002) reported that adsorption of organic esters (e.g. biodiesel) over polar surfaces such as those of silica and Nafion resins, is negligible. Yori et al. (2007) studied the reversible adsorption of glycerol from biodiesel and reported that silica has a great capacity for glycerol removal, its saturation capacity being 0.13 g of glycerol per gram of adsorbent. When operated in packed beds, for a glycerol concentration of 0.11−0.25% typical of biodiesel streams issuing from gravity settling tanks, an effluent limit of $C/C^0=0.01$ and an entrance velocity of 11 cm min^{-1}, a 2 m high silica bed with 1/8" beads would have a net processing capacity of 0.01−0.02 m^{3}biodiesel kg$_{silica}^{-1}$. Much of the good performance of silica is related to the favorable thermodynamics of adsorption, since glycerol-silica displays an almost irreversible, square isotherm (Yori, 2008).

4.3 Soaps, salts and metals

Soaps are produced by the reaction of FFAs during the first steps of caustic refining of the fatty feedstock or by the reaction of the remaining FFAs with alkaline homogeneous catalysts in the transesterification reactor. These reactions lead to the formation of estearates, oleates, palmitates, etc. of sodium and potassium, that are amphiphilic substances that bring phase separation and plugging problems downstream the reactor. Other salts of sodium or potassium come from the neutralization of acid homogeneous catalysts in the acid-catalyzed process. These inorganic salts lead to corrosion in lines and vessels and they must also be completely eliminated in the final biodiesel product because of quality issues.

Metals are minor components in all oils as they are present as oligoelements in highly specialized molecules such as chlorophylls (magnesium) and porphyrins (magnesium, iron, manganese). Other sources of metals are the contamination from iron and copper surfaces during the process of oil extraction or biodiesel production. Certain metals, such as cobalt, manganese and chromium, but particularly iron and copper, exhibit a prooxidant effect in oil. The manifestations of oxidation are flavor, color and odor deterioration. Copper is perhaps the most active catalyst, exhibiting noticeable oxidation properties at levels as low as 0.005 ppm (Flider & Orthoefer, 1981). Though flavor, color and odor deterioration are probably not an issue for biodiesel, oxidation stability is indeed required.

For soaps, salts and metals, adsorption on silica adsorbents seems the most suitable means of removal (Welsh et al., 1990). Clays offer only a small adsorption capacity for soaps and an almost null capacity for metals.

4.4 Free fatty acids

FFAs have negligible values in biodiesel produced by the alkaline method. Depending on the efficiency of esterification they can be present in non-negligible amounts in biodiesel produced by the acid-catalyzed method or the supercritical method. Manuale et al. (2011) reacted different feedstocks with acidities ranging from 0.08 to 23.6% and found that the esterification with supercritical methanol (280 °C, 20=methanol-to-oil ratio) reduced the FFA content to 1-2.5% after 1 h and 0.4-0.6% after 1.5 h of reaction time. Reduction of the FFA content to values lower than those of the international norms can be done by washing. Adsorption however can prove simple, robust and efficient. For these application silicas are found to be superior than other adsorbents in both bleaching capacity and bleaching rate.

Adsorbent	Adsorbent conc., mass %	Bleaching time, min	Adsorption capacity, $g_{FFA}\ g_{ads}^{-1}$
Virgin activated carbon	5	720	6.0
Mg doped activated carbon	5	720	5.0
Diatomaceous earth	1	30	10.1
Silica gel	0.36	90	140.0

Table 3. Adsorbents capacity for FFA removal from biodiesel (Manuale et al., 2011).

5. Adsorption

In the last years there has been a great progress in adsorbent design and cyclic adsorption process developments, thus making adsorption an important separation tool (King, 1980). Adsorption is usually performed in columns packed with adsorbent but it can also be performed in stirred tanks with the adsorbent in suspension. The latter are usually known as bleachers since their most common application is the bleaching of edible oils with clays.

The high separating power of the chromatographic effect, achieved in adsorbent-packed columns, is a unique advantage of adsorption as compared to other separation processes. The high separating power is caused by the continuous contact and equilibration between the fluid and sorbent phases. If no diffusion limitations are considered, each contact is equivalent to an equilibrium stage (theoretical plate) and several hundreds or more of such equilibrium stages can be achieved within a short column. Adsorption is thus ideally suited for purification applications and difficult separations.

The adsorptive separation is achieved by one of three mechanisms: adsorption equilibrium, steric effect and kinetic effect. Most processes, especially those in solid-liquid phase, operate with the principle of adsorption equilibrium and hence they are called equilibrium separation processes. In this processes the amount of adsorbate retained is primarily determined by the thermodynamic adsorbate-adsorbent activity with little regard to mass transfer phenomena. The steric effect derives from the molecular sieving properties of zeolites and other molecular sieves and can be taken as an extreme case of adsorption controlled by mass transfer phenomena. In this case either small or properly shaped molecules can diffuse into the adsorbent while other molecules are partially or totally excluded. Typical examples are the separation of linear and branched alkanes (Silva et al., 2000) or the dehydration of aqueous ethanol (Teo & Ruthven, 1986), both performed using molecular sieves. Kinetic separation is achieved by virtue of the differences in diffusion rates of different molecules. This kind of separation is mostly found in gas-gas separation as in the separation of the component gases of air (Ruthven & Farooq, 1990).

In the case of the biodiesel feedstock and product, the low elution rates in the packed columns makes the dynamic separation (kinetic effect) of no use for a practical separation. In the case of the steric effect this is expected to work fine for molecules differing widely in size and this could be the case for molecules of the organic and polar phases normally found at the outlet of the transesterification reactors. Triglycerides, diglycerides, monoglycerides, free fatty acids and fatty acid methyl esters have high molecular weights and long acyl chains and they are the main components of the organic phase. On the other side glycerol, water and methanol have small molecular sizes and could be retained in packed beds containing suitable adsorbents. Because of their relative high vapor pressure, water and methanol need a relatively few number of thoretical plates to be separated from the organic phase by distillation/evaporation (Zhang et al., 2003) and this is indeed the preferred method of water and methanol removal. However some reports on the use of hygroscopic adsorbents for biodiesel drying can be found (Lastella, 2005). Removal of glycerol from biodiesel using adsorbents has already been proved but only equilibrium adsorption on open pore adsorbents has been tried (Yori et al., 2007; Mazzieri et al., 2008). The use of the steric effect in the adsorption of water on zeolites has however been proposed for the drying of the methanol to be recycled to the biodiesel process (McDonald, 2001).

This leaves equilibrium adsorption as the main principle behind the adsorption refining of biodiesel and makes the adsorption isotherm as the main piece of information for the accurate design and scale-up of adsorption units. In this sense, though a lot of information is available for adsorption of impurities from plant oils (biodiesel feestock) in relation to bleaching with clays (Hussin et al., 2011) or silicas (Rossi et al., 2003) only scarce information for purification of biodiesel by adsorption has been published (Manuale et al., 2011; Schmitt Faccini et al., 2011; Vasques, 2009; Mazzieri et al., 2008).

6. Adsorption isotherms

The function that describes the relation between the amount of adsorbate on the solid and its liquid-phase concentration is called adsorption equilibrium isotherm. Different functions can be used to describe this equilibrium. The Langmuir-type isotherm remains to be the most widely used for practical applications (Eq. 1).

$$\theta = \frac{q^*}{q_m} = \frac{K_L C^*}{1 + K_L C^*} \tag{1}$$

Only liquid phase applications will be discussed in this review and therefore also only liquid phase isotherms. The constant K_L is called Langmuir constant. C^* and q^* are the equilibrium values of the bulk concentration of the adsorbate in the liquid phase and the concentration in the solid phase. θ is the fractional coverage of the surface and q_m the maximum or saturation load. At low pressures or in dilute solutions, the Langmuir isotherm reduces to a linear form, or Henry's law form (Eqs. 2-3).

$$\theta = K_L C * \tag{2}$$

$$q = K_L C * q_m = HC * \tag{3}$$

All isotherms should reduce to the Henry's law form at extreme dilution. Since high dilution is the condition for many systems that need to be purified to extremely small amounts of certain impurities, the Henry's constant becomes the most important factor for purification. Both K_L and H are proportional to the exponential of the heat of adsorption $(-\Delta H)$. For physical adsorption, ΔH is proportional to the bond energy between the adsorbate molecule and the adsorbent surface. Thus bond energy becomes critical for purification. Strong bonds are typical of adsorbate-adsorbent systems in ultrapurification.

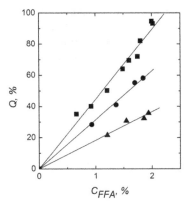

Fig. 2. Adsorption isotherm for silica TrySil 3000 at three different temperatures (Manuale, 2011). 70 °C (■), H=44.6. 90 °C (●), H=31.4. 110 °C (▲), H=18.3. ΔH=-5.7 kCal mol⁻¹.

Zeldowitsch (1934) and previously Freundlich (1906) supplied an equation that is widely used to describe the data for heterogeneous adsorbents (Eq. 4).

$$q^* = K_F \, C *^{1/n} \tag{4}$$

In this formula q^* and C^* are the equilibrium adsorbate concentrations in the solid and fluid phase, respectively. Zeldowitsch obtained this formula assuming an exponentially decaying function of site density with respect to ΔH, while Freundlich proposed it on an empirical basis. Freundlich's isotherm is customarily used to express the equilibrium concentration of metals and colorant bodies (chlorophylls, carotenes, etc.) in oils (Liu et al., 2008; Toro-Vázquez & Proctor, 1996) and is expected that it should also be convenient for the same adsorbates in biodiesel precursor oils and fats. In the case of the biodiesel product probably the fit of the data of adsorption of some impurities could also be good, but in this case the oil has already been refined before entering the transesterification reactor and in so diluted condition the Henry's linear isotherm could better apply.

To avoid indefinite increase in adsorption with concentration, the so-called Langmuir-Freundlich isotherm is sometimes proposed (Sips, 1948) (Eq. 5). This isotherm can be derived from the Langmuir isotherm by assuming each adsorbate molecule occupies n sites. It can also be considered as the Langmuir isotherm on nonuniform surfaces.

$$(q*/q_m) = \frac{K_{LF}\, C*^{1/n}}{1 + K_{LF}\, C*^{1/n}}.$$
(5)

Langmuir's formula has been successfully used to express the adsorption of glycerides from biodiesel over silica gel (Mazzieri et al., 2008). In the case of free fatty acids (FFAs) Nawar and Han (1985) also concluded that the Langmuir isotherm was followed by octanoic acid adsorption on silica. The better adjustment of free fatty acid (oleic, linoleic, etc.) adsorption over several solids by the Langmuir model (in comparison to Freundlich's) has also been reported by Proctor and Palaniappan (1990) and Cren et al. (2005, 2010).

The Langmuir and Langmuir-Freundlich isotherms for adsorption of single components are readily extended to an n-component mixture to yield the extended multicomponent Langmuir isotherm (Yang, 1997) (Eq. 6) and the so-called loading ratio correlation (LRC) (Yon & Turnock, 1971) (Eq. 7). In these equations it is assumed that the area occupied by one molecule is not affected by the presence of other species on the surface. This is not thermodynamically consistent but the equations remain nonetheless useful for design.

$$(q_i*/q_{m,i}) = \frac{K_{L,i}\, C_i*}{1 + \sum K_{L,i}\, C_i*}$$
(6)

$$(q_i*/q_{m,i}) = \frac{K_{LF,i}\, C_i*^{1/n}}{1 + \sum K_{LF,i}\, C_i*^{1/n}}$$
(7)

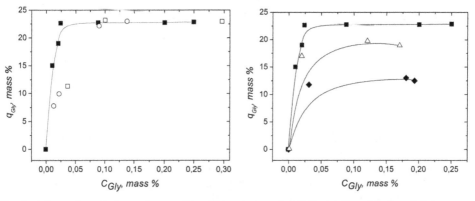

Fig. 3. Adsorption of glycerol over silica (Mazzieri et al., 2008). (■) Pure biodiesel. (□) Biodiesel spiked with water (944 ppm). (○) Biodiesel spiked with soap (270 ppm). (Δ) Biodiesel spiked with MG (7500 ppm). (♦) Biodiesel spiked with methanol (12000 ppm).

Mazzieri et al. (2008) used the multicomponent Langmuir isotherm to express the simultaneous adsorption of glycerol and monoglycerides. They found that adsorption of glycerol is not influenced by the presence of small amounts of water and soaps. Conversely the presence of MGs and/or methanol lowers the adsorption capacity of glycerol because of the competition of MGs for the same adsorption sites.

7. Mass transfer kinetics and models for adsorption in the liquid phase

It is generally recognized that transfer of adsorbates from the bulk of a liquid occurs in two stages. First molecules diffuse through the laminar film of fluid surrounding the particles and then they diffuse inside the pore structure of the particle. Most authors assume that the concentration gradient of any species along the film is linear and that the mass transfer to the adsorbent surface is proportional to the so-called film coefficient, k_f (Eq. 8). In this equation, q is the adsorbent concentration on the solid particle, r_p is the particle radius and ρ_p is the average density of the particle. C is the concentration of the adsorbate in the bulk of the fluid and C_s the value of adsorbate concentration on the surface. k_f is often predicted with the help of generalized, dimensionless correlations of the Sherwood (Sh) number that correlate with the Reynolds (Re) and Schmidt (Sc) numbers and the geometry of the systems. The most popular is that due to Wakao and Funazkri (1978) (Eq. 9).

$$\frac{\partial q}{\partial t} = \left(\frac{3k_f}{r_p \rho_p} \right)(C - C_s) \tag{8}$$

$$Sh = \frac{2r_p k_f}{D_m} = \left(2.0 + 1.1 Sc^{\frac{1}{3}} Re^{0.6} \right) \tag{9}$$

$$Sc = \frac{\mu}{D_M \rho} \tag{10}$$

In the case of the homogeneous surface diffusion model (HSDM) the equation of mass transport inside the pellet it that of uniform Fickian diffusion in spherical coordinates (Eq. 11). Sometimes this model is modified for system in which the diffusivity is seemingly not constant. The most common modification is to write the surface diffusivity, D_s, as a linear function of the radius, thus yielding the so-called proportional diffusivity model (PDM). A detailed inspection of the available surface diffusivity data indicates that surface diffusivity is similar but expectedly smaller than molecular diffusivity, D_M. Some values of D_M are presented in Table 4.

$$\frac{\partial q}{\partial t} = D_s \left(\frac{\partial^2 q}{\partial r^2} + \frac{2}{r} \frac{\partial q}{\partial r} \right) \tag{11}$$

In the case of fatty substances there is not much reported data on the values of surface diffusivity. Yang et al. (1974) found that stearic acid had a surface diffusivity on alumina of about 10^{-9}-10^{-11} m^{2}s^{-1} depending on the hydration degree of the alumina. Allara and Nuzzo (1985) reported values of D_s of 10^{-10}-10^{-11} for different alkanoic acids on alumina.

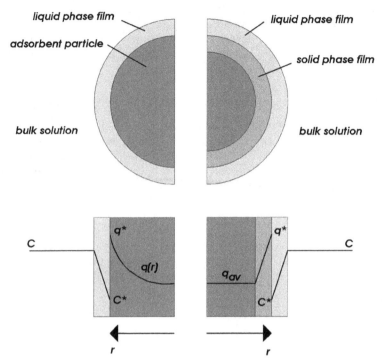

Fig. 4. Homogeneous surface diffusion (left) and linear driving force (right) models.

Molecule	T, °C	Solvent	D_M, m^2 s^{-1}	Reference
Stearic acid	130	Nut oil	4.2×10^{-10}	Smits (1976)
Oleic acid	130	Nut oil	3.7×10^{-10}	Smits (1976)
Monoolein	25	Water	1.3×10^{-10}	Geil et al. (2000)
Triolein, Tristearin	70	Triolein, tristearin	$1\text{-}2 \times 10^{-10}$	Callaghan & Jolley (1980)
Sodium oleate	25	Sodium oleate	3.3×10^{-10}	Gajanan et al. (1973)
Sodium palmitate		Sodium palmitate	4.8×10^{-10}	Gajanan et al. (1973)
Glycerol	130	Biodiesel	6.18×10^{-9}	Kimmel (2004)

Table 4. Values of molecular diffusivity of several biodiesel impurities.

$$\frac{\partial q}{\partial t} = K_{LDF}(q^* - q_{av}) \tag{12}$$

In the case of the linear driving force model (LDFM) all mass transfer resistances are grouped together to give a simple relation (Eq. 12). q_{av} is the average adsorbate load on the pellet and is obtained by the time-integration of the adsorbate flux. q^* is related to C^*, through the equilibrium isotherm. It must be noted that in this formulation $q_s=q^*$ and $C_s=C^*$, indicating that the surface is considered to be in equilibrium. In the case of adsorption for refining of biodiesel, the LDF approximation has been used to model the adsorption of free fatty acids over silicas (Manuale, 2011). FFA adsorption was found to be rather slow despite the small diameter of the particles used (74 microns). This was addressed to the dominance of the intraparticle mass transfer resistance. This resistance was attributed to a working mechanism of surface diffusion with a diffusivity value of about 10^{-15} m^2 s^{-1}. The system could be modeled by a LDFM with an overall coefficient of mass transfer, K_{LDF}=0.013-0.035 min^{-1} (see Table 5). These values compare well with those obtained for the adsorption of sodium oleate over magnetite, 0.002-0.03 min^{-1} (Roonasi et al., 2010).

Adsorbent	T, °C	K_{LDF}, min^{-1}	Adsorbent	T, °C	K_{LDF}, min^{-1}
Silica TrySil 3000	70	0.035	Silica TrySil 300B	70	0.032
	90	0.019		90	0.022
	110	0.013		110	0.018

Table 5. Values of the LDF overall mass transfer coefficient for the silica adsorption of free fatty acids from biodiesel at different temperatures (Manuale, 2011).

The authors provided a further insight into the internal structure of the LDF kinetic parameter by making use of the estimation originally proposed by Ruthven et al. (1994) for gas phase adsorption (Eq. 13). D_s is the intrapellet surface diffusivity and ε is the porosity of the pellet. The additivity of the intrapellet diffusion time (τ_D) and the film transfer time (τ_f) to give the total characteristic time ($1/K_{LDF}$=τ_{total}) is sometimes questioned because of the large difference between them. In the case of the adsorption of oleic acid from biodiesel it was shown that $\tau_f\approx0.07$ seconds (estimated) and $\tau_{total}\approx1700$ seconds (experimental) indicating that the silica-FFA system is strongly dominated by intrapellet diffusion (Manuale, 2011).

$$\frac{1}{K_{LDF}} = \frac{r_p}{3k_f} + \frac{r_p^2}{15\varepsilon D_s} = \tau_f + \tau_D \qquad (13)$$

The LDF model was first proposed by Glueckauf and Coates (1947) as an "approximation" to mass transfer phenomena in adsorption processes in gas phase but has been found to be highly useful to model adsorption in packed beds because it is simple, analytical, and physically consistent. For example, it has been used to accurately describe highly dynamic PSA cycles in gas separation processes (Mendes et al., 2001). Yet, a difference is sometimes found in the isothermal batch uptake curves on adsorbent particles obtained by the LDFM and the more rigorous HSDM. The LDF approximation has also been reported to introduce some error when the fractional uptake approaches unity (Hills, 1986). In practice however saturation values might never be approached because adsorption capacity is severely decreased due to unfavourable thermodynamics in the saturation range. The precision of LFDM can be also improved by using higher order LDF models (Álvarez-Ramírez et al., 2005).

8. Experimental breakthrough curves

Breakthrough curve. It is the "S" shaped curve that results when the effluent adsorbate concentration is plotted against time or volume. It can be constructed for full scale or pilot testing. The breakthrough point is the point on the breakthrough curve where the effluent adsorbate concentration reaches its maximum allowable concentration, which often corresponds to the treatment goal, usually based on regulatory or risk based numbers.

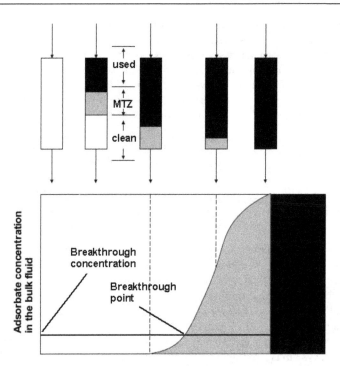

Fig. 5. Adsorption colum zones. Relation to breakthrough curve.

Mass Transfer Zone. The mass transfer zone (*MTZ*) is the area within the adsorbate bed where adsorbate is actually being adsorbed on the adsorbent. The *MTZ* typically moves from the influent end toward the effluent end of the adsorbent bed during operation. That is, as the adsorbent near the influent becomes saturated (spent) with adsorbate, the zone of active adsorption moves toward the effluent end of the bed where the adsorbate is not yet saturated. The *MTZ* is generally a band, between the spent adsorbent and the fresh adsorbent, where adsorbate is removed and the dissolved adsorbate concentration ranges from C^o (influent) to C^e (effluent). The length of the *MTZ* can be defined as L_{MTZ}. When $L_{MTZ}=L$ (bed length), it becomes the theoretical minimum bed depth necessary to obtain the desired removal. As adsorption capacity is used up in the initial *MTZ*, the *MTZ* advances down the bed until the adsorbate begins to appear in the effluent. The concentration gradually increases until it equals the influent concentration. In cases where there are some very strongly adsorbed components, in addition to a mixture of less strongly adsorbed components, the effluent concentration rarely reaches the influent concentration because only the components with the faster rate of movement are in the breakthrough curve. Adsorption capacity is influenced by many factors, such as flow rate, temperature, and pH (liquid phase). The adsorption column can be considered exhausted when C^e equals 95 to 100% of C^o.

9. Model equations for flow in packed beds

We should start by writing the general equation for flow inside a packed bed, isothermal, and with no radial gradients (Eqs. 14-17). In these equations, u is the interstitial velocity

($u=U/\varepsilon_B$), where U is the empty bed space velocity and ε_B is the bed porosity. The last three equations are the "clean bed" initial condition and the Danckwertz boundary conditions for a closed system.

$$\frac{\partial C}{\partial t} - D_L \frac{\partial^2 C}{\partial z^2} + \frac{\partial(uC)}{\partial z} + \frac{1-\varepsilon_B}{\varepsilon_B}\rho_p \frac{\partial q}{\partial t} = 0 \tag{14}$$

$$C(0,t) = C^0 \tag{15}$$

$$\frac{\partial C}{\partial z} = 0, \quad z = L \tag{16}$$

$$C(z,0) = 0 \tag{17}$$

In order to solve a specific problem of adsorption, mass transfer kinetics equations must be added, such as those of the HSDM or LDFM. The film equation is customarily replaced in the general equation of flow along the bed (Eq. 14) and thus the total system is reduced. The system still remains rather complex and in most instances can only be solved numerically. For faster convergence and accuracy special methods can be used, such as orthogonal collocation, the Galerkin method, or finite element methods. The general solution of the system is a set of points of C as a function of z, t and r. Often much of this information is not necessary and only the fluid bulk concentration at the bed outlet as a function of time, i.e. the "breakthrough" curve, is reported.

In order to obtain analytical breakthrough curves some simplifications can be made. For example the first implication of a high intrapellet diffusion resistance in liquid-solid systems (as in biodiesel refining) is that the Biot number that represents the ratio of the liquid-to-solid phase mass transfer rate, takes very high values. In Biot's equation (Eq. 18), q^0 is the equilibrium solid-phase concentration corresponding to the influent concentration C^0 and r_p is the particle radius. The film resistance in high Bi systems can be disregarded; their breakthrough curves being highly symmetrical. Experimental symmetrical curves have indeed been found for the adsorption of glycerol over packed beds of silica (Fig. 6).

$$Bi = \frac{k_f r_p C^0}{D_s \rho_p q^0} \tag{18}$$

Another simplification is related to the longitudinal dispersion term in Eq. 14. D_L is usually calculated together with the film coefficient k_f by using the Wakao & Funazkri (1978) correlations for the mass transfer in packed beds of spherical particles (Eqs. 9 and 19). Due to the dependence of Sc on the molecular diffusivity, the value of D_L is dominated by D_M. The importance of D_L in systems of biodiesel flowing in packed bed adsorbers could be disregarded in attention to the value of the axial Péclet number (Eq. 22), since $Pe > 100$ in these systems. For very big Pe numbers the regime is that of plug flow (no backmixing) and when Pe is very small the backmixing is maximum and the flow equations are reduced to the equation of the perfectly mixed reactor (Busto et al., 2006).

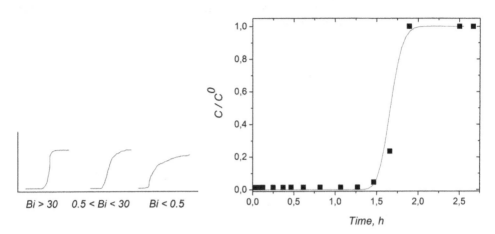

Fig. 6. Left: appearance of breakthrough curves as a function of the Biot number. Right: breakthrough curve for glycerol adsorption over silica (Yori et al., 2007).

$$\frac{D_L}{2\mu r_p} = \frac{20}{Re\,Sc} + 0.5 \tag{19}$$

$$Pe = \frac{uL}{D_M} \tag{20}$$

Another degree of complexity is posed by the nature of the isotherm equilibrium equation. Langmuir and Langmuir-Freundlich formulae are highly linear and propagate this non-linearity to the whole system. However some simplifications can be done depending on the strength of the affinity of the adsorbate for the surface and the range of concentration of the adsorbate of practical interest.

Sigrist et al. (2011) have indicated that Langmuir type isotherms for systems with high adsorbate/solid affinity can be approximated by an irreversible "square" isotherm ($q=q_m$), while systems in the high dilution regime can be represented by the linear Henry's adsorption isotherm. Combining the linear isotherm or the square isotherm with the equations for flow and mass transfer along the bed, inside the pellet and through the film, analytical expressions for the breakthrough curve of biodiesel impurities over silica beds can be found (Table 6) (Yori et al., 2007).

For the square isotherm, the Weber and Chakravorti (1974) model is depicted in equations 21-25. A square, flat isotherm curve yields a narrow MTZ, meaning that impurities are adsorbed at a constant capacity over a relatively wide range of equilibrium concentrations. Given an adequate capacity, adsorbents exhibiting this type of isotherm will be very cost effective, and the adsorber design will be simplified owing to a shorter MTZ. Weber and Chakravorti took a further advantage of this kind of isotherm and simplified the intrapellet mass transfer resolution by supposing that the classical "unreacted core" model applied, i.e., that the surface layers could be considered as completely saturated and that a mass front diffused towards the "unreacted core".

Isotherm	Film resistance	Intrapellet resistance	Adsorption	Biodiesel system	References
Linear	Yes	Fick, CD	Reversible	FFA-silica	Rasmusson & Neretnieks (1980)
Square	Yes	Fick, CD	Irreversible	Glycerol-silica	Weber & Chakravorti (1974)

Table 6. Breakthrough models for square and linear isotherms. CD: constant diffusivity.

$$\tau - N_p = \frac{15}{\sqrt{3}} \tan^{-1} \left[\frac{2(1-Q)^{1/3}}{\sqrt{3}} + 1 \right] - \frac{15}{2} \ln \left[1 + (1-Q)^{1/3} + (1-Q)^{2/3} \right] +$$

$$+ 2.5 - \frac{5\pi}{2\sqrt{3}} + \left(\frac{N_p}{N_f} \right) \ln (Q+1) \tag{21}$$

$$\tau = \left[\frac{15\,\varepsilon\,D_s}{r_p^2} \right] \left[\frac{C^0}{q_m} \right] (t - z/u). \tag{22}$$

$$N_p = \left[\frac{15\,\varepsilon\,D_S}{r_p^2} \right] \left[\frac{1-\varepsilon_B}{\varepsilon_B} \right] \left(\frac{z}{u} \right) \tag{23}$$

$$N_f = k_f \left[\frac{1-\varepsilon}{\varepsilon} \right] \left(\frac{3z}{u\,r_p} \right) \tag{24}$$

$$Q = \frac{q}{q_s} = \frac{C}{C^0} \tag{25}$$

τ is the dimensionless time variable, Q is the fractional uptake, N_p is the pore diffusion dimensionless parameter and N_f is the film dimensionless parameter. The constant pattern condition is fulfilled in most of the span of the breakthrough experiments ($\tau > 5/2 + N_p/N_f$) except in the initial region when the pattern is developing. The simplified expression for dominant pore diffusion (high Bi) can be obtained by setting (N_p/N_f)=0.

For glycerol adsorption over silica Yori et al. (2007) provided a sensitivity study based on Weber and Chakravorti's model. These results are plotted in Figures 7 and 8. The influence of the pellet diameter (d_p) can be visualized in Figure 7 at two concentration scales. For small diameter (1 mm) the saturation and breakthrough points practically coincide and the traveling MTZ is almost a concentration step. For higher diameters the increase in the time of diffusion of glycerol inside the particles produces a stretching of the mass front and a more sigmoidal curve appears. The breakthrough point was defined as $C/C^0=0.01$ because for common C^0 values (0.1-0.25% glycerol in the feed) lowering the glycerol content to the quality standards for biodiesel (0.002%) demands that C/C^0 at the outlet is equal or lower than 1% the value of the feed. The results indicate that for a 3 mm pellet diameter the breakthrough time is reduced from 13 h to 8 h and that for a 4 mm pellet diameter this value is further reduced to 4.5, i.e. almost one third the saturation time. It can be inferred that the pellet diameter has a strong influence on the processing capacity of the silica bed. Small diameters though convenient from this point of view are not practical. d_p is usually 3-6 mm in industrial adsorbers in order to reduce the pressure drop and the attrition in the bed.

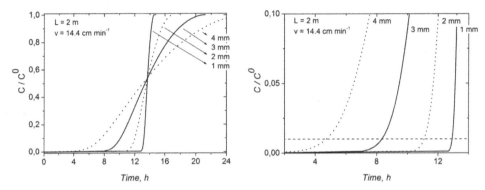

Fig. 7. Adsorption of glycerol from biodiesel. Breakthrough curves as a function of pellet diameter (d_p). Breakthrough condition C/C^0=0.01, L=2 m, U=14.4 cm min^{-1}.

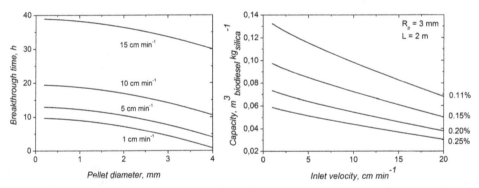

Fig. 8. Adsorption of glycerol from biodiesel. Left: breakthrough time as a function of U and d_p (L=2 m, U=14.4 cm min^{-1}). Right: influence of U and C^0 on the processing capacity (d_p=3 mm, L=2 m).

The combined influence of pellet diameter and inlet velocity on the breakthrough time is depicted in Figure 8 (left). The breakthrough time seems to depend on d_p^{-n} (n>0) and also on U^{-n} (n>0). This means that longer breakthrough times are got at smaller pellet diameters and smaller feed velocities. The processing capacity per unit kg of silica is displayed in Figure 8 (right) as a function of d_p and the inlet velocity, U^0. When U^0 goes to zero the bed capacity equals q_m, and decreases almost linearly when increasing U^0. For a typical solid-liquid velocity of 5 cm min^{-1} the capacity decreases at higher glycerol concentration, but the silica bed is used more efficiently because the relative MTZ size is reduced.

$$y(\tau)=\frac{1}{2}\mu^0\left(1+erf\left\{\frac{(\ln(\tau)-\mu)}{(\sigma\sqrt{2})}\right\}\right) \qquad (26)$$

$$Bi^* = \frac{k_f r_p}{HD_s \rho_s} \qquad (27)$$

$$\Theta = \frac{\varepsilon_B L D_s}{u \, r_p^2} \qquad (28)$$

The breakthrough curve for the linear isotherm model is depicted in equations (26-28). This is the Q-LND (quasi log normal distribution) approximation of Xiu et al. (1997) and Li et al. (2004), of the general solution of Rasmusson and Neretnieks (1980). This approximation is known to be valid in systems of high Bi. y is the adimensional adsorbate concentration in the fluid phase, τ is the adimensional time, μ and σ parameters are functions of the Péclet number (Pe), the modified Biot number (Bi^*) and the time parameter (Θ).

10. Experimental scale-up of adsorption columns

The Rapid Small Scale Column Test (RSSCT) was developed to predict the adsorption of organic compounds in activated carbon adsorbers (Crittenden et al., 1991). The test is based upon dimensionless scaling of hydraulic conditions and mass transport processes. In the RSSCT, a small column (SC) loaded with an adsorbent ground to small particle sizes is used to simulate the performance of a large column (LC) in a pilot or full scale system. Because of the similarity of mass transfer processes and hydrodynamic characteristics between the two columns, the breakthrough curves are expected to be the same. Due to its small size, the RSSCT requires a fraction of the time and liquid volume compared to pilot columns and can thus be advantageously used to simulate the performance of the large column at a fraction of the cost (Cummings & Summers, 1994; Knappe et al., 1997). As such, RSSCTs have emerged as a common tool in the selection of adsorbent type and process parameters.

Parameters of the large column are selected in the range recommended by the adsorbent vendor. The RSSCT is then scaled down from the large column. Based on the results of the RSSCT, the designer develops detailed design and operational parameters. The selection and determination of the following parameters is required:

- Mean particle size: the designer must find an adequate mesh size, 100-140, 140-170, 170-200, etc., that can be used to successfully simulate the large column. Too small particles can however lead to high pressure losses and pumping problems.
- Internal diameter (ID) of column: 10-50 mm ID columns are preferred to keep all other column dimensions small and more important, to reduce the amount of time and eluate used. The $d_{SC}/d_{p,SC}$ should be higher than 50 to keep wall effects negligible.

RSSCT scaling equations have been developed with both constant (CD) and proportional (PD) diffusivity assumptions. The two approaches differ if D_s values are independent (for CD) or a linear function (for PD) of the particle diameter, d_p. Equations 29-30 can be used to select the small column (SC) RSSCT parameters based upon a larger column (LC) that is being simulated. t is the time span of the experiment for a common outlet concentration. For CD and PD scenarios the values for X are zero and one, respectively. Additional X values have been suggested based upon non-linear relationships between d_p and D_s.

$$\frac{EBCT_{SC}}{EBCT_{LC}} = \left(\frac{d_{p,SC}}{d_{p,LC}}\right)^{2-x} = \frac{t_{SC}}{t_{LC}} \qquad (29)$$

$$X = \log\left(\frac{d_{p,SC}}{d_{p,LC}}\right) / \log\left(\frac{D_{s,SC}}{D_{s,LC}}\right) \tag{30}$$

- The spatial or interstitial velocities (U, u) are scaled based on the relation written in Eq. 31. However, this equation will result in a high interstitial velocity of water in the small column, and hence, high head loss. Crittenden (1991) recommended that a lower velocity in the small column be chosen, as long as the effect of dispersion in the small column does not become dominant over other mass transport processes. This limitation requires the $Re_{SC}Sc$ value remain in the range of 200-200,000, which is the mechanical dispersion range.

$$\frac{u_{SC}}{u_{LC}} = \left(\frac{d_{p,\ LC}}{d_{p,\ SC}}\right) \tag{31}$$

Variable	Small column	Large column
d_p	0.3 mm	3 mm
EBCT	105 s	2.9 h
U	2.4 mm s^{-1}	0.24 mm s^{-1}
L	25 cm	2.5 m
t_{run}	3 days	300 days

Table 7. Variables for a scaled-down constant diffusivity RSSCT packed with silica gel for adsorption of glycerol. Values for the small column taken from Yori et al. (2007).

In the case of biodiesel, no results of RSSCTs designed for scale-up purposes have been published so far, though some tests in small columns have been published (Yori et al., 2007). The validity of RSSCTs holds anyway. In this sense one first step for their use for scale-up purposes would be to determine the kind of D_S-d_p relation that holds, since it is unknown whether CD or PD approaches must be used. In order to show the usefulness of the technique, a procedure of comparison between a biodiesel large column adsorber and a scaled down laboratory column is made in Table 7.

11. Advantages of adsorption in biodiesel refining

As pointed out by McDonald (2001), Nakayama & Tsuto (2004), D'Ippolito et al. (2007), Özgül-Yücel & Turkay (2001) and others, the principal advantage of the use of adsorbers in biodiesel refining is that of reducing the amount of wastewater and sparing the cost of other more expensive operations such as water washing and centrifugation. For big refiners that can afford the cost of setting up a water treatment plant the problem of the amount of wastewater might not be an issue but this can be extremely important for small refiners.

In the common industrial practice water-washing is used to remove the remaining amounts of glycerol and dissolved catalyst, and also the amphiphilic soaps, MGs and DGs. Theoretically speaking if water-washing is used to remove glycerol and dissolved catalyst only, large amounts of water should not be required. However in the presence of MGs and DGs the addition of a small amount of water to the oil phase results in the formation of an emulsion upon stirring. Particularly when this operation is performed at a low temperature separation of the aqueous phase from the emulsion becomes difficult. In order to prevent the formation of such an emulsion in the conventional water-washing practice a large amount of water must be used. Karaosmanoglu et al. (1996) concluded that a minimum of 3-5 grams of water per gram of biodiesel at 50 °C were needed to efficiently remove the impurities of the fuel (3000-5000 litres of water per Ton of biodiesel). These numbers should be considered typical of once-through water-washing operations but are not representative of closed-loop water washing schemes. Accurate numbers are included in Table 8.

It has been suggested that the methanol removal step needed for succesful adsorption be performed before glycerol separation and under vacuum conditions (D'Ippolito et al., 2007; Bournay et al., 2005). The data in Table 8 suggests that the best operation of dry refining is that with cyclic reversible adsorption of glycerol/glycerides in twin packed beds, as early suggested (D'Ippolito et al., 2007).

	Lurgi	*Crown Iron*	*Dry*
Glycerol removal from biodiesel	Water wash column	Water mixer/settler	Packed bed, bleacher
Methanol removal from biodiesel	Water wash column	Steam Stripper	Vacuum flash drum
Methanol removal from wash water	Rectifier column	Rectifier column	Not needed
Final polishing by bleaching	n.a.	Yes	Not needed
Wash water consumption	200 kg Ton_{bio}^{-1}	200 kg Ton_{bio}^{-1}	None
Adsorbent consumption	n.a.	n.a.	11 kg Ton_{bio}^{-1} (bleacher) < 1 kg Ton_{bio}^{-1} (cyclic bed)

Table 8. Comparison of unit operations for two alkali-catalyzed processes (Lurgi, 2011; Crown Iron Works, 2011; Anderson et al., 2003) and a process with a "dry" step of adsorption of glycerol and glycerides (Manuale et al., 2011). Adsorbent comsumption calculated for glycerol removal only (0.15% in raw biodiesel) (Yori et al., 2007).

Other advantages of adsorption are the low capital investment (provided common adsorbents are used), the absence of moving parts, the simplicity and robustness of operation. Possible drawbacks are the need for disposal and replacement of the spent adsorbent in the case of the use of bleaching tanks.

12. Adsorbers operation

12.1 Bleaching tanks

Manuale et al. (2011) used bleaching silicas for the removal of FFA in biodiesel in a series of tests in a stirred tank reactor under varying temperature and pressure conditions (70 and 110 °C, 760 and 160 mm_{Hg}). Their results confirm the pattern already seen in the case of the silica refining of edible oils. For the same adsorbent and in the presence of vacuum the influence of temperature is low. For example for TriSyl 3000 in vacuo, after 90 min, and from a similar initial acidity level (1.5%), the adsorbate load at two different temperatures is: $q_{70\,°C}$=99.3%, $q_{110\,°C}$=75.0%. Similarly, for TriSyl 300B, 90 min bleaching time, 1.7-1.9% initial acidity: $q_{70\,°C}$=82.0, $q_{110\,°C}$= 69.0%. The trend is clear. Higher temperatures lead to lower adsorption capacities. This is related to the fact that adsorption is exothermal and thus adsorption equilibrium is favored at low temperatures. In the absence of vacuum, adsorption is very low, one order of magnitude the value at 160 mm$_{Hg}$. Water adsorption reportedly inhibits the diffusion and adsorption inside the pore network of the silicas. At 90 °C or higher temperatures water desorption from an adsorbent dipped in oil can only proceed to a non-negligible extent in the presence of vacuum. Therefore if the adsorbent is not previously dehydrated, dehydration occurs simultaneously with adsorption during the bleaching experiment. In some cases the release of water from the silica goes directly into the biodiesel phase and the water content of the oil phase is increased.

These results indicate that surface diffusion of FFA over several adsorbents is very slow and the limiting step of the whole adsorption process. This leads to two negative consequences: (i) if a high level of FFA removal and a short bleaching time is required then big amounts of adsorbent must be used and these adsorbents are only partially used; (ii) if a total utilization of the adsorbent is desired, unconveniently high bleaching times must be used.

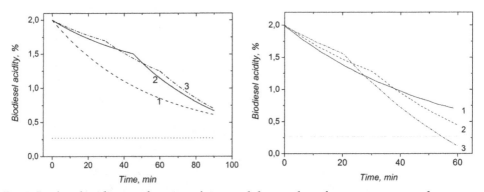

Fig. 9. Biodiesel acidity as a function of time and the number of countercurrent tank bleaching steps (1, 2 and 3) (Manuale, 2011). Adsorbent load=2%, initial biodiesel acidity= 2%, K_{LDF}= 0.0188 min^{-1}. Left: Linear adsorption (Henry's law, H=37.6). Right: Irreversible adsorption (square isotherm). Dotted line: FFA European standard EN 14214, FFA limit.

Manuale et al. (2011) discussed the conditions for total total adsorbent utilization and for quasi complete FFA removal. They used the LDF model with both linear and square isotherms. They tested by simulation the use of serial cocurrent and countercurrent bleachers in order to assess their bleaching performance. The results are presented in Figure 9. In the case of the linear adsorption isotherm the use of countercurrent bleachers does not lead to a reduction of the adsorbent consumption. An effective reduction only occurs when the isotherm is square. These conclusions hold independently of the number of serial bleachers. Therefore when adsorption is strong and irreversible, spent adsorbents can be used advantageously to bleach streams highly contaminated while fresh adsorbents can be used to polish bleach the most lean streams. In the case of the linear isotherm the modulation of the adsorption capacity results in an operation that depends only on the bleaching time (all traces in Figure 9-left coincide at the end of the bleaching cycle).

Figure 10 is a plot of $q(t)$ for a train of countercurrent beds packed with adsorbents having a linear isotherm. The results show that all traces for the multistep operation are practically parallel to the 1-bleacher trace. Hence the adsorption capacity q is only a function of the "total" bleaching time. No benefits can then be got from the multi-tank countercurrent bleaching operation. The only possibility of multiple units is that of parallel bleaching tanks working long times (e.g. 2 h) in order to increase the adsorbent usage.

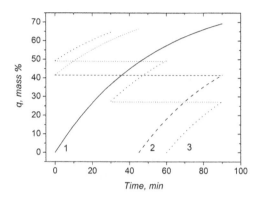

Fig. 10. Adsorbent load as a function of time and the number of countercurrent bleaching steps (1, 2 and 3) (Manuale, 2011). Process conditions as in Figure 9.

12.2 Packed beds

Lead-lag operation. Most liquid phase packed beds are operated in series. This means passing all of the flow through one column bed, a lead column, and then passing flow through another similar sized column bed, the lag vessel. This method offers several advantages over a single column. The series configuration allows the maximum use of the adsorbent throughout the entire bed. This assumes that the MTZ is contained within a single properly sized packed bed. By placing two or more columns in series, the MTZ is allowed to pass completely through the first (lead) bed as the leading edge of the MTZ migrates into the second (lag) bed. By allowing this to happen, the maximum contaminant concentration is allowed to come into contact with adsorption sites in the lead vessel that require a greater concentration gradient to hold additional contamination. When the MTZ exits the lead vessel, that vessel is then exhausted, and requires change out with virgin or regenerated adsorbent. Even though the adsorption capacity of the lead vessel is exhausted, treatment continues in the lag vessel. Then, during change out, the lead vessel is taken off-line and the lag vessel is placed in the lead position. The former lead vessel is then replenished with adsorbent and then becomes the lag vessel and brought on-line. Further insights on the operation of serial and parallel adsorbers can be found elsewhere (Sigrist et al., 2011).

Regeneration. For the removal of glycerol and to a lower extent of MGs and DGs, the methanol concentration in the fluid is important. Methanol adversely affects the adsorption capacity because it increases the activity of glycerol and glycerides in the liquid phase. This was studied by Yori et al. (2007) with the method disclosed by Condoret (1997) and Bellot (2001). The method is based on the knowledge of the curves describing the variation in the glycerol activity with respect to its concentration, established separately for each phase (solid and liquid). Henry's constants were obtained from the slope of the isotherms in the diluted range using the UNIFAC method for calculating the liquid phase activity coefficients. The results are shown in Fig. 11 and indicate that for all practical purposes the adsorption of glycerol over silica is null at high methanol concentrations.

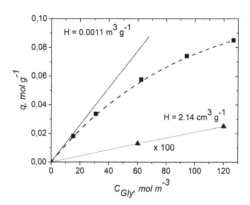

Fig. 11. Silica adsorption isotherms for the Gly-FAME (squares) and Gly-MeOH (triangles) systems. H values calculated from the slope of the traces. Yori et al. (2007).

The elution of 4 bed volumes of methanol through the exhausted packed bed is reported to restore the adsorbent capacity. Elimination of the adsorbed methanol from the silica was done by blowing nitrogen through the bed but could be performed using any other gas. Elimination of the solvent produced a decrease of the bed temperature because methanol evaporation needs 1104 J g⁻¹. This translates to 200 kJ kg$_{silica}$⁻¹ for the fully saturated silica and hence provisions should be made in order to maintain the bed temperature and prevent biodiesel flow problems at unconvenient low temperatures. In this sense flushing the bed with a hot gas seems the most suitable means for desorbing methanol.

$$(H/H^0) = e^{-\frac{\Delta H}{R}\left[\frac{1}{T_1} - \frac{1}{T_2}\right]}$$
(32)

The thus recommended way of regenerating the silica bed seems superior to other means used for regeneration of adsorbent packed beds, notably the thermal swing. A thermal swing with purified hot biodiesel could be used to regenerate the bed. Manuale et al. (2011) found that the silica adsorption of oleic acid from biodiesel has a heat of adsorption of -5.7 KCal mol⁻¹. This is similar to reported values for similar systems (Sari & Ipýldak, 2006). In order to decrease the adsorption capacity 100 times ($H/H°$=0.01, Eq. 32) the thermal swing should be ΔT=480 °C. For mild regenerations with $H/H°$=0.1 and $H/H°$=0.25, the required thermal swings are still high, ΔT=200 °C and ΔT=140 °C. The results indicate that though for the silica-FFA system adsorption is weak enough to yield a linear isotherm, the heat of adsorption is too high and discourages the use of a thermal swing for regeneration.

13. Conclusions

Adsorption is a robust and reliable operation for the refining of biodiesel and its feedstocks. Hydrophillic adsorbents seem the best choice, because most of the undesired impurities are polar. In this sense silicas offer a high saturation capacity (10-15%) for glycerol and glycerides, and enough affinity for soaps, FFA, metals and salts.
One advantage of adsorption units for the removal of glycerol, glycerides, soaps, phosphatides and metals from biodiesel and its feedstocks, is the reduction in wastewater

effluents and the sparing of washing, oil-water separation and wastewater treatment units. Other advantages are small capital expenditure, robustness and easiness of operation.

Cost-effective means for the scale-up of packed bed adsorbers for biodiesel refining seem to be accurate models for flow and adsorption and scaled-down RSCCTs. Accurate models for flow and adsorption can be solved in their full complexity only with the aid of numerical calculations but analytical solutions for rapid design and sensitivity analysis can be got using approximations, such as the use of square and linear isotherms and LDF models. Further approximations can be obtained for low Biot and high axial Péclet numbers.

The operation of adsorbers should minimize the consumption of adsorbent. From this point of view countercurrent bleaching tank arrays should be used but this mode of operation cannot be exploited in the case of adsorbents with linear isotherms. In the case of packed bed adsorbers common lead-lag setups of 2 or more serial columns are recommended.

14. Acknowledgements

This work was financed with the support of Universidad Nacional del Litoral (Grant PI-60-298, CAI+D 2009) and CONICET (National Research Council of Argentina).

15. References

Allara, D.L. & Nuzzo, R.G. (1985) Spontaneously organized molecular assemblies. 1. Formation, dynamics, and physical properties of n-alkanoic acids adsorbed from solution on an oxidized aluminum surface. *Langmuir*, Vol 45, No 1, pp. 45-52.

Álvarez-Ramírez, J., Fernández-Anaya, G., Valdés-Parada, F.J. & Ochoa-Tapia, J.A. (2005). Physical Consistency of Generalized Linear Driving Force Models for Adsorption in a Particle. *Industrial & Engineering Chemistry Research*, Vol. 44, No 17, pp. 6776-6783, ISSN 08885885.

Anderson, D., Masterson, D., McDonald, B. & Sullivan, L. (2003). Industrial Biodiesel Plant Design and Engineering: Practical Experience. *Chemistry and Technology Conference, Session Seven: Renewable Energy Management, International Palm Oil Conference (PIPOC)*, 24-28 August 2003, Putrajaya Marriot Hotel, Putrajaya, Malaysia.

Bellot, J.C., Choisnard, L., Castillo, E. & Marty, A. (2001). Combining solvent engineering and thermodynamic modeling to enhance selectivity during monoglyceride synthesis by lipase-catalyzed esterification. *Enz. Microb. Technol.*, Vol. 28, No 4-5, pp. 362-369, ISSN 0141-0229.

Bondioli, P., Gasparoli, A., Bella, L. D. & Tagliabue, S. (2002). Evaluation of biodiesel storage stability using reference methods. *Eur. J. Lipid Sci. Tech.*, Vol. 104, No 12, pp. 777-784, ISSN 1438-9312.

Bournay, L., Casanave, D., Delfort, B., Hillion, G. & Chodorge, J. (2005). New heterogeneous process for biodiesel production: A way to improve the quality and the value of the crude glycerin produced by biodiesel plants. *Catalysis Today*, Vol. 106, No. 1-4, pp. 190-192, ISSN 0920-5861.

Busto, M., D'Ippolito, S.A., Yori, J.C., Iturria, M.E., Pieck, C.L., Grau, J.M. & Vera, C.R. (2006). Influence of the axial dispersion on the performance of tubular reactors during the non-catalytic supercriticaltransesterification of triglycerides. *Energy & Fuels*, Vol 20, No 6, pp. 2642-2647, ISSN 0887-0624.

Callaghan, P.T., Jolley & K.W. (1980). Translational motion in the liquid phases of tristearin, triolein and trilinolein. *Chem. Phys. Lipids*, Vol. 27, No 1, pp. 49-56, ISSN 1016-0009.

Chang, Y., van Gerpen, J.H., Lee, I., Johnson, L.A., Hammond, E.G. & Marley, S.J. (1996). Fuel properties and emissions of soybean oil esters as diesel fuel. *Journal of the American Oil Chemists' Society*, Vol. 73, No. 11, pp. 1549-1555, ISSN 0003-021X.

Condoret, J.S., Vankan, S., Joulia, X. & Marty, A. (1997). Prediction of water adsorption curves for heterogeneous biocatalysis in organic and supercritical solvents. *Chem. Eng. Sci.*, Vol. 52, pp. 213-220, ISSN 0009-2509.

Cren, E.C. & Meirelles, A.J.A. (2005). Adsorption Isotherms for Oleic Acid Removal from Ethanol + Water Solutions Using the Strong Anion-Exchange Resin Amberlyst A26 OH. *J. Chem. Eng. Data*, Vol. 50, No 5, pp 1529–1534

Cren, E.C., Morelli, A.C., Sanches, T., Rodrigues, C.E. & Meirelles, A.J.A. (2010). Adsorption Isotherms for Removal of Linoleic Acid from Ethanolic Solutions Using the Strong Anion Exchange Resin Amberlyst A26 OH. *J. Chem. Eng. Data*, Vol. 55, No. 7, pp 2563–2566.

Crittenden, J.C., Reddy, P.S., Arora, H., Trynoski, J., Hand, D.W., Perram, D.L. & Summers, R.S. (1991). Predicting GAC Performance With Rapid Small-Scale Column Tests. *Journal American Water Works Association*, Vol. 83, No. 1, pp. 77-87, ISSN 0003150X.

Crown Iron Works (2011). Biodiesel Plants Technical Brochure.

Cummings, L., & Summers, R.S. (1994). Using RSSCTs to Predict Field-Scale GAC Control of dbp Formation. *Journal American Water Works Association*, Vol. 86, No. 6, pp 88-97, ISSN 0003-150X.

Darnoko, D. & Cheryan, M. (2000). Kinetics of palm oil transesterification in a batch reactor. *JAOCS*, Vol. 77, No 12, pp 1263-1267, ISSN 0003-021X .

Díaz, Z. & Miller, J.H. (1990). Removal of polar impurities from diesel and jet fuel. United States Patent 4,912,873.

D'Ippolito, S.A., Yori, J.C., Iturria, M.E., Pieck, C.L. & Vera, C.R. (2007). Analysis of a Two-Step, Noncatalytic, Supercritical Biodiesel Production Process with Heat Recovery, *Energy & Fuels*, Vol. 21, No 1, pp 339-346, ISSN 0887-0624.

Dunn, R. O., Shockley, M.W. & Bagby, M.O. (1996). Improving the low-temperature properties of alternative diesel fuels: Vegetable oil-derived methyl esters. *JAOCS*, Vol. 73, pp. 1719-1728.

Du Plessis, L. M.; de Villiers, J. B. M. & van der Walt, W. H. (1985). Stability studies on methyl and ethyl fatty acid esters of sunflowerseed oil. *JAOCS*, Vol. 62, No 4, pp 748-752, ISSN 0003-021X .

Falk, O. & Meyer-Pittroff, R. (2004). The effect of fatty acid composition on biodiesel oxidative stability. *European Journal of Lipid Science and Technology*, Vol 106, No 12, pp 837-843, ISSN 1438-7697

Farhan, F.M., Rahmati, H. & Ghazi-Moghaddam, G. (1988). Variations of trace metal content of edible oils and fats during refining processes. *JAOCS*, Vol. 65, No 12, pp. 1961-1962, ISSN 0003-021X .

Flider, F.J. & Orthoefer, F.T. (1981). Metals in soybean oil. *JAOCS*, Vol. 58, No. 3, 270-272, ISSN 0003-021X.

Foletto, E.L., Colazzo, G.C., Volzone, C. & Porto, L.M. (2011). Sunflower oil bleaching by adsorption onto acid-activated bentonite. *Brazilian Journal of Chemical Engineering*, Vol. 28, No. 1, pp. 169-174, ISSN 0104-6632.

Freedman, B., Butterfield R. O. & Pryde E. H. (1986). Transesterification kinetics of soybean oil. *JAOCS*, Vol. 63, No 10, pp 1375-1380, ISSN 0003-021X.

Freundlich, H. (1906). Über die Adsorption in Lösungen. *Z. Phys. Chem., Vol* 57, pp 385-470.

Geil, B., Feiweier, T., Pospiech, E.M., Eisenblätter, J., Fujara, F. & Winter, R. (2000). Relating structure and translational dynamics in aqueous dispersions of monoolein. *Chemistry and Physics of Lipids*. Vol. 106, No 2, pp. 115-126, ISSN 0009-3084 .

Glueckauf, E. & Coates, J. (1947). Theory of chromatography. Part IV. The influence of incomplete equilibrium on the front boundary of chromatograms and on the effectiveness of separation. *J. Chern. Soc.*, pp. 1315-1321.

Goto, V., Sasaki, T. & Takagi, K. (2004). Method and apparatus for preparing fatty acid esters. US Patent 6,812,359.

Grace Davison, TriSyl silica for edible oil refining, Technical Brochure, 2005.

Griffin, R.P. & Dranoff, J.S. (1963). The rate of glycerol absorption by ion exchange resins. *AIChE Journal*, Vol. 9, No 2, pp, 283-&, ISSN 0001-1541.

He, H., Sun, S., Wang, T. & Zhu, S. (2007). Transesterification kinetics of soybean oil for production of biodiesel in supercritical methanol. *JAOCS*, Vol. 84, No. 4 pp. 399-404, ISSN 0003-021X.

Hills, J.H. (1986). An investigation of the linear driving force approximation to diffusion in spherical particles. *Chemical Engineering Science*, Vol. 41, No 11, pp. 2779-2785.

Hussin, F., Kheireddine, M. & Wan Daud, W.M.A. (2011). Textural characteristics, surface chemistry and activation of bleaching earth: A review. *Chemical Engineering Journal*, Vol. 170, No 1, pp. 90-106, ISSN 1385-8947.

Ito, E., Rob van Veen, J.A. & Ng, F.T.T. (2006). On novel processes for removing sulphur from refinery streams. *Catalysis Today*, Vol. 116, No. 4, pp 446–460, ISSN 0920-5861.

Karaosmanoglu, F., Cigizoglu, K.B., Tuter, M. & Ertekin, S. (1996). Investigation of the Refining Step of Biodiesel Production. *Energy & Fuels*, Vol. 10, No. 4, pp. 890-895, ISSN 0887-0624.

Kimmel, T., Doctoral Thesis, Technical University of Berlin, 2004.

King, C. J. (1980). Separation Processes, 2nd Ed., McGraw-Hill, ISBN 9780070346123, New York, USA.

Knappe, D.R.U., Snoeyink, V.L., Roche, P., Prados, M.J. & Bourbigot, M.M. (1997). The effect of preloading on rapid small-scale column test predictions of atrazine removal by GAC adsorbers. *Water Research*, Vol. 31, No 11, pp. 2899-2909, ISSN 0043-1354.

Lastella, J.M. (2005). System for removal of methanol from crude biodiesel fuel. United States Patent Application 20050188607.

Li, P., Xiu, G. & Rodrigues, A.E. (2004). Modeling breakthrough and elution curves in fixed beds of inert core adsorbents: analytical and approximate solutions. *Chem. Eng. Sci.*, Vol. 59, No. 15, pp. 3091-3103, ISSN 0009-2509.

Liu, Y., Huang, J. & Wang, X. (2008). Adsorption Isotherms for Bleaching Soybean Oil with Activated Attapulgite. *JAOCS*, Vol. 85, No 10, pp. 979-984, ISSN 0003-021X .

Lotero, E., Liu, Y., Lopez, D.E., Suwannakarn, K., Bruce, D.A. & Goodwin, Jr., J.G. (2005). Synthesis of Biodiesel via Acid Catalysis. *Industrial & Engineering Chemistry Research*, Vol. 44, No 14, pp. 5353-5363, ISSN 0888-5885.

Lurgi GmbH (2011). Biodiesel Plants Technical Bulletin. Frankfurt am Main, Germany.

Manuale, D.L., Mazzieri, V.A., Torres, G., Vera, C.R. & Yori, J.C. (2011). Non-catalytic biodiesel process with adsorption-based refining. *Fuel*, Vol. 90, No 3, pp. 1188-1196, ISSN 0016-2361.

Manuale, D.L. Tesis Doctoral, Universidad Nacional del Litoral, Santa Fe, Argentina, 2011.

Mazzieri, V.A., Vera, C.R. & Yori, J.C. (2008) Adsorptive properties of silica gel for biodiesel refining. *Energy & Fuels*, Vol. 22, No 6, pp 4281-4284, ISSN 0887-0624.

Mendes, A.M.M., Costa, C.A.V. & Rodrigues, A.E. (2001). Oxygen separation from air by PSA: modelling and experimental results. Part I: isothermal operation. *Separation and Purification Technology*, Vol. 24, No1-2, pp. 173-188, ISSN 1383-5866.

McDonald, W.M. (2001). Process for dry synthesis and continuous separation of a fatty acid methyl ester reaction product. US Patent 6,262,285.

Moser, B.R., Haas, M.J., Winkler, J.K., Jackson, M.A., Erhan, S.Z. & List, G.R. (2007). Evaluation of partially hydrogenated methyl esters of soybean oil as biodiesel. *Eur. J. Lipid Sci. Technol.*, Vol. 109, No 1, pp 17–24, ISSN 1438-7697.

Nakayama, M. & Tsuto, K. (2004). Method of producing fatty acid alkyl ester for diesel fuel oil. European Patent EP 1 477 549.

Nawar, W.W. & Han, L.B. (1985), Thermal oxidation of lipids in monolayers I. The nature of binding on silica, *JAOCS*, Vol. 62, No 11, pp 1596-1598, ISSN 0003-021X.

Negi, D.S., Sobotka, F., Kimmel, T., Wozny, G. & Schomäcker, R. (2006). Liquid-liquid phase equilibrium in glycerol-methanol-methyl oleate and glycerol-monoolein-methyl oleate ternary systems. *Ind. Eng. Chem. Res.*, Vol. 45, No 10, pp. 3693-3696, ISSN 0888-5885.

Nijhuis, T.A., Beers, A.E.W., Kapteijn, F. & Moulijn, J.A. (2002). Water removal by reactive stripping for a solid-acid catalyzed esterification in a monolithic reactor. *Chem. Eng. Sci.*, Vol. 57, No 9, pp. 1627-1632, ISSN 0009-2509.

Noureddini, H., Dailey, W.R. & Hunt, B.A. (1998). Production of ethers of glycerol from crude glycerol the by-product of biodiesel production. *Advances in Environmental Research*, Vol. 2, No 2, pp 232-243.

Noureddini, H. & Zhu, D. (1997). Kinetics of transesterification of soybean oil. *JAOCS*, Vol. 74, No 11, pp. 1457-1463, ISSN 0003-021X .

Özgül-Yücel, S. & Turkay, S. (2001). Purification of FAME by rice hull ash adsorption. *JAOCS*, Vol. 80, No 4, pp. 373-376, ISSN 0003-021X

Portilho, M.F., Vieira, J.A.V., Zotin, J.L. & Lima, M.S.S. (2008). Method for the production of biodiesel from vegetable oils and fats, using heterogeneous catalysts. United States Patent Application 20080295393.

Proctor, A. & Palaniappan, S. (1990). Adsorption of soy oil free fatty acids by rice hull ash, JAOCS, Vol. 67, No 1, pp 15-17, ISSN 0003-021X.

Rasmusson, A. & Neretnieks, I. (1980). Exact solution of a model for diffusion in particles and longitudinal dispersion in packed beds. *AIChE J.*, Vol. 26, No. 3, pp. 686–690, ISSN 0001-1541.

Refaat, A.A. (2009). Correlation between the chemical structure of biodiesel and its physical properties. *Int. J. Environ. Sci. Tech.*, vol. 6, No 4, pp. 677-694, ISSN 1735-1472.

Romig, C. & Spataru, A., (1996), Emissions and engine performance from blends of soya and canola methyl esters with ARB #2 diesel in a DDC 6V92TA MUI engine, *Bioresource Technology*, Vol. 56, No 1, pp 25-34, ISSN 0960-8524.

Rossi, M., Gianazza, M., Alamprese, C. & Stanga, F. (2003). The role of bleaching clays and synthetic silica in palm oil physical refining. *Food Chemistry*, Vol. 82, No 2, pp. 291-296, ISSN 0308-8146.

Roonasi, P., Yang, X. & Holmgren, A. (2010). Competition between Na oleate and Na silicate for a silicate/oleate modified magnetite surface studied by in situ ATR-FTIR spectroscopy. *J. Colloid. Interface Sci.* 343, No 2, pp. 546-552, ISSN 0021-9797.

Ruthven, D.M. & Farooq, S. (1990). *Gas Separation & Purification*, Vol. 4, No 3, pp 141-148, ISSN 0950-4214 .

Ruthven, D.M., Farooq, S. & Knaebel, K.S. (1994). Pressure Swing Adsorption. VCH Publishers, New York, ISBN 1560815175.

Saka, S. & Minami, E. (2009). Biodiesel production with supercritical fluid technologies. *Handbook of Plant-Based Biofuels*, CRC Press, ISBN: 978-0-7890-3874-6.

Saka, S. & Kusdiana, D. (2001). Biodiesel fuel from rapeseed oil as prepared in supercritical methanol. *Fuel*, Vol. 80, No 2, pp. 225-231, ISSN 0016-2361 .

Sari, A. & Ipýldak, Ö. (2006). Adsorption properties of stearic acid onto untreated kaolinite. *Bull. Chem. Soc. Ethiop.*, Vol. 20, No 2, pp. 259-67, ISSN 1011-3924 .

Schmitt Faccini, M., Espinosa da Cunha, M., Aranda Moraes, M.S., Krause, L.C., Maniquel, M.C., Rodrigues, M.R.A., Benvenutti, E.V. & Caramão, E.B. (2011). Dry washing in biodiesel purification: a comparative study of adsorbents. *J. Braz. Chem. Soc.*, Vol. 22, No.3, pp 558-563, ISSN 0103-5053.

Sigrist, M., Beldoménico, H., Tarifa, E.E., Pieck, C.L. & Vera, C.R. (2011). Modelling diffusion and adsorption of As species in Fe/GAC adsorbent beds. *J. Chem. Tech. Biotech.*, DOI: 10.1002/jctb.2638.

Silva, J.A.C., Da Silva, F.A. & Rodrigues, A.E. (2000), Separation of n/iso paraffins by PSA, *Separation and Purification Technology*, Vol. 20, No 1, pp. 97–110, ISSN 1383-5866 .

Sips, R. (1948). On the structure of a catalyst surface, J. Chem. Phys., Vol. 16, No 5, pp 490-495, ISSN 0021-9606

Smits, G. (1976). Measurement of the diffusion coefficient of free fatty acid in groundnut oil by the capillary-cell method. *JAOCS*, Vol. 53, No 4, pp. 122-124, ISSN 0003-021X .

Tang, H.; Wang, A.; Salley, S. O. & Simon, K. Y., (2008). The Effect of Natural and Synthetic Antioxidants on the Oxidative Stability of Biodiesel. *JAOCS*, Vol. 85, No 4, pp 373-382, ISSN 0003-021X.

Teo, W.K. & Ruthven, D.M. (1986). Adsorption of water from aqueous ethanol using 3-A° molecular sieves. *Ind. Eng. Chem. Process Des. Dev.*, Vol. 25, No. 1, pp. 17–21, ISSN 0196-4305.

Thompson, J. C., Peterson, C. L., Reece, D. L. & Beck, S. M. (1998). Two-year storage study with methyl and ethyl esters of rapeseed. *Am. Soc. Agric. Eng.*, Vol. 41, No 4, pp 931-939, ISSN 0001-2351 .

Toro-Vázquez, J.F. & Proctor A. (1996). The freundlich isotherm in studying adsorption in oil processing. *JAOCS*, Vol. 73, No 12, pp. 1627-1633, ISSN : 0003-021X.

van Dalen, J.P. & van Putte, K.P.A.M. (1992). Adsorptive refining of liquid vegetable oils. *Congrès DGF-Meeting No 48*, Vol. 94, NS2, pp. 567–570, ISSN 0931-5985 .

van Gerpen, J.H., Hammond, E.G., Johnson, L.A., Marley, S.J., Yu, L., Lee, I. & Monyem, A. (1996). Determining the Influence of Contaminants on Biodiesel Properties. *Final Report for the Iowa Soybean Promotion Board*, Iowa State University, July 31.

Vasques, E. (2009). Adsorção de glicerol, mono e diglicerídeos presentes no biodísel produzido a partir do óleo de soja. Master Thesis, UFPR.

Wakao, N. & Funazkri, T. (1978) Effect of fluid dispersion coefficients on particle-to-fluid mass transfer coefficients in packed beds. Correlation of Sherwood numbers. *Chem. Eng. Sci.*, Vol. 33, No 10, pp 1375-1384, ISSN 0009-2509 .

Wang, W.G., Lyons, D.W., Clark, N.N., Gautam, M. & Norton, P.M. (2000). Emissions from nine heavy trucks fueled by diesel and biodiesel blend without engine modification. *Environ. Sci. Technol.*, Vol. 34, No 6, pp 933-939, ISSN 0013-936X .

Warabi, Y., Kusdiana, D. & Saka, S. (2004). Reactivity of triglycerides and fatty acids of rapeseed oil in supercritical alcohols. *Biores. Technol.*, Vol. 91, No 3, pp. 283-287, ISSN 0960-8524 .

Weber, T.W. & Chakravorti, R.K. (1974). Pore and solid diffusion models for fixed-bed adsorbers. *AIChE J.*, Vol. 20, No 2, pp. 228-237, ISSN 0001-1541 .

Welsh, W.A., Bogdanor, J.M. & Toenebohen, G.J. (1990). Silica refining of oils and fats. In *Proceedings of the World Conference on Edible Fats and Oils Processing*, p. 89, Maastricht, The Netherlands, Oct. 1-7, 1989. Edited by D.R. Erickson.

Xiu, G.H., Nitta, T., Li, P. & Jin, G. (1997). Breakthrough curves for fixed-bed adsorbers: quasi-lognormal distribution approximation. *AIChE J.*, Vol. 43, No 4, pp. 979–985, ISSN 0001-1541.

Yang, R.T. (1997). Gas Separation by Adsorption Processes. Butterworth, 1987, reprinted by Imperial College Press, London.

Yang, R.T., Fenn, J.B. & Haller, G.L. (1974). Surface diffusion of stearic acid on aluminum oxide. *AIChE J.*, Vol. 20, No 4, pp 735-742,ISSN 0001-1541 .

Yon, C.M. & Turnock, P.H. (1971). Multicomponent adsorption equilibria on molecular sieves. *AIChE Symp. Ser.*, Vol. 67, No. 117, pp. 75-83.

Yori, J.C., D'Amato, M.A., Grau, J.M., Pieck, C.L. & Vera, C.R. (2006). Depression of the cloud point of biodiesel by reaction over solid acids. *Energy & Fuels*, Vol. 20, No 6, pp 2721-2726, ISSN 0887-0624.

Yori, J. C., D'Ippolito, S. A., Pieck, C. L. & Vera, C. R. (2007). Deglycerolization of Biodiesel Streams by Adsorption Over Silica Beds. *Energy & Fuels*, Vol. 21, No 1, pp 347-353, ISSN 0887-0624.

Yori, J.C. (2008). Doctoral Thesis, UNED, Madrid, Spain.

Zeldowitsch, J. (1934). Adsorption site energy distribution. *Acta Physicochim, URSS*, Vol. 1, pp 961–973.

Zhang, Y., Dubé, M.A., McLean, D.D. & Kates, M. (2003). Biodiesel production from waste cooking oil: 1. Process design and technological assessment. *Bioresource Technology*, Vol. 89, No 1, pp 1–16, ISSN 0960-8524 .

Zhou, W. & Boocock, D.G.B. (2006). Phase distributions of alcohol, glycerol, and catalyst in the transesterification of soybean oil. *JAOCS*, Vol. 83, No 12, pp 1047-1052, ISSN 0003-021X.

The Immobilized Lipases in Biodiesel Production

Margarita Stoytcheva[1], Gisela Montero[1],
Lydia Toscano[1], Velizar Gochev[2] and Benjamin Valdez[1]
[1]Universidad Autónoma de Baja California, Instituto de Ingeniería,
[2]Plovdiv University "Paisii Hilendarski,
[1]México
[2]Bulgaria

1. Introduction

The leading standard setting organization ASTM International, formerly ASTM (American Society of Testing and Materials), defines biodiesel as a fuel comprised of mono-alkyl esters of long chain fatty acids (ASTM D6751). It is usually manufactured by triglycerides transesterification with methanol or ethanol in the presence of a catalyst, according to the following reaction:

$$
\begin{array}{ccccccc}
& O & & & & O & \\
& \| & & & & \| & \\
CH_2\text{-}O\text{-}C\text{-}R_1 & & & & R_1\text{-}C\text{-}O\text{-}R & & CH_2\text{-}OH \\
| & O & & & & O & | \\
| & \| & & & \xrightarrow{\text{catalyst}} & \| & | \\
CH\text{-}O\text{-}C\text{-}R_2 & + & 3ROH & & R_2\text{-}C\text{-}O\text{-}R & + & CH\text{-}OH \\
| & O & & \text{alcohol} & & O & | \\
| & \| & & & & \| & | \\
CH_2\text{-}O\text{-}C\text{-}R_1 & & & & R_3\text{-}C\text{-}O\text{-}R & & CH_2\text{-}OH \\
\end{array}
$$

| triglycerides (vegetable oil, animal fat) | mono-alkyl esters (biodiesel) | glycerol |

Recently, biodiesel production by lipase catalyzed transesterification has been suggested as a promising alternative to the conventional chemical catalysis, in spite of the high conversion and reaction rates of the latter (Akoh et al., 2007; Bajaj et al., 2010; Bisen et al., 2010; Demirbas, 2009; Fjerbaek et al., 2009; Fukuda et al., 2001, 2009; Ghaly et al., 2010; Helwani et al., 2009; Jegannathan & Abang, 2008; Man Xi Ao et al., 2009; Marchetti et al., 2007; Ranganathan et al., 2008; Robles-Medina et al., 2009; Semwal et al., 2011). The enzymatic process enables eliminating the drawbacks of the alkali- or acid-catalyzed transesterification, namely: product contamination, wastewater release, free fatty acids and water interferences, and difficult glycerol recovery. Nevertheless, the commercialization of the lipase-catalyzed biodiesel synthesis remains problematic, because of the cost of the enzyme: approximately 1000 USD per kg of Novozym 435 lipase. Therefore, the implementation of strategies, such as enzyme immobilization, for the development of economic and effective enzyme based technologies for biodiesel production is of crucial importance. Enzyme immobilization ensures several issues: repetitive and continuous use of the enzyme and its stabilization, localization of the interaction, prevention of product contamination, reduction of effluent problems and material handling, and effective control of the reaction parameters (D'Souza, 1982). All these aspects are reflected on the production cost.

The present review is intended to provide an overview on the use of immobilized lipases in biodiesel production, the techniques applied for enzyme immobilization, and the factors affecting the process.

2. Lipases mode of action and classification

Lipases (EC 3.1.1.3 triacylglycerol acylhydrolase) represent a group of water soluble enzymes that originally catalyze the hydrolysis of ester bonds in water insoluble lipid substrates, acting at the interface between the aqueous and the organic phases. This unique heterogeneous reaction is feasible because of: (i) the specific lipases molecule 3D structure consisting of three domains: "contact domain", responsible for distinguishing of substrate surface, "hydrophobic" domain, responsible for extracting of one substrate molecule and its association with the "functional" domain, and "functional" domain, containing the catalytic triad Ser, Hys and Asp/Glu; (ii) the transition from closed to open conformation in the presence of the lipidic phase (Guncheva & Zhiryakova 2011; Panalotov & Verger, 2000). Enzymatic action of lipases on the substrate is a result of a nucleophilic attack on the carbonyl carbon atom from ester groups. Some lipases are also able to catalyze the processes of esterification, interesterification, transesterification, acidolysis, amynolysis and may show enantioselective properties (Hasan et al., 2009). The mechanism of the lipase catalyzed transesterification of triglycerides with an alcohol to produce biodiesel (Fjerbaek et al., 2009) could be presented by the following sequence of reactions:

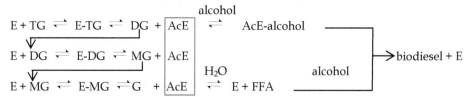

with: E-enzyme; TG-triglyceride; DT-diglyceride; MG-monoglyceride; G-glycerol; AcE-acylated enzyme; FFA-free fatty acid.

According to the origin lipases are plant, animal and microbial. The mostly used lipases in biodiesel production are of bacterial and fungal origin, such as: *Candida antarctica* (Novozym 435), *Candida Rugosa* (Lipase AY), *Pseudomonas cepacia* (Lipase PS), *Pseudomonas fluorescens* (lipase AK), *Pseudomonas aeruginosa*, and *Thermomyces lanuginose* (Lipozime TL), among other. The catalytic properties and potential applications of *Bacillus* lipases are extensively reviewed by Guncheva & Zhiryakova (2011). Among the available lipase producing microorganisms, filamentous fungi belonging to various species of genera *Aspergillus* (Adinarayana et al., 2004; Karanam, & Medicherla, 2008), *Rhizopus* (Hiol et al., 2000; Shukla

& Gupta, 2007), *Penicillium* (Chahinian et al, 2000; Lima et al., 2003; Vardanega et al., 2010), and *Trichoderma* (Kashmiri et al., 2006; Rajesh et al., 2010) are described as the most prospective lipase producers. Only microbial lipases are a matter of practical interest for biodiesel production, because only microbial lipases are produced in industrial scale. The application of the microbial lipases, all together with their immobilization which allows the regeneration and the reuse of the enzyme preparation in several working cycles reduces the production costs, and respectively the final cost of biodiesel. A review on microbial lipase production with emphasis on lipase engineering and use of mathematical models for process improvement and control is provided by Treichel et al. (2010).

Some fungi cultured by the authors as powerful producers of lipase of use in biodiesel synthesis are shown in Fig. 1.

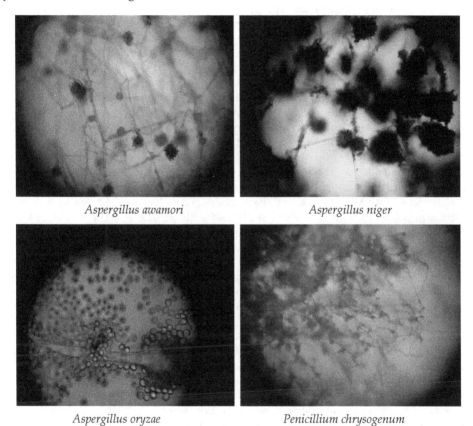

Aspergillus awamori   *Aspergillus niger*

Aspergillus oryzae   *Penicillium chrysogenum*

Fig. 1. Filamentous fungi belonging to various species of genera *Aspergillus* and *Penicillium*, considered as prospective lipase producers.

3. The immobilized lipases in biodiesel production

The term "enzyme immobilization" was defined at the first Enzyme Engineering Conference held at Hennicker, NH, USA, in 1971. It describes "enzymes physically confined at or localized in a certain region of space with retention of their catalytic activity and which can be used repeatedly and continuously" (Powel, 1996). It is considered that lipase immobilization induces the enzyme conformational change required to enable the free access of substrate to the active centre. Especially, hydrophobic supports allow the adsorption of the open form of the lipases via interfacial activation, mimicking the lipophilic substrate (Ahn et al., 2010; Rodrigues & Fernandez-Lafuente, 2010; Salis et al., 2008; Séverac et al., 2011).

The revision of the literature covering the period 2005-2011 demonstrates that a large variety of matrices have been used for lipases immobilization, and that the main methods applied include adsorption, entrapment and/or encapsulation, and covalent attachment.

3.1 Lipases immobilization by adsorption

Physical adsorption is considered as the simplest method for enzyme immobilization. Enzyme fixation is performed through hydrogen bonds, salt linkages, and Van der Waal's forces. The process is carried out in mild conditions, without or with minimum support activation and clean up procedures application, and in the absence of additional reagents. Thus, it is economic and allows preserving enzyme activity and specificity. The chemical composition of the carrier, the molar ratio of hydrophilic to hydrophobic groups, as well as the particle size and the surface area determine the amount of enzyme bound and the enzyme behaviour after immobilization. Some of the most commonly used carriers for lipases immobilization by adsorption are listed in Table 1.

Data shown in Table 1 indicate that among the variety of lipase immobilization supports, Accurel has found a large application. Accurel is the trade name of a group of macroporous polymers. As carriers for lipase immobilization are used the polypropylene based hydrophobic Accurel MP (MP1000 with particle size below 1500 μm, Accurel MP1001 with particle size below 1000 μm and Accurel MP1004 with particle size below 400 μm), and Accurel EP-100. On the most hydrophobic support tested, Accurel MP1001, no glycerol adsorption was observed (Séverac et al., 2011). It has been demonstrated (Salis et al., 2009) comparing the catalytic efficiencies (activity/loading) of eight lipases, that they show a different level of adaptation to the support. Immobilized *Pseudomonas fluorescens* lipase is the most active biocatalyst, followed by immobilized *Pseudomonas cepacia* lipase. The other lipases tested (from *Rhizopus oryzae, Candida rugosa, Mucor javanicus, Penicillium roqueforti, Aspergillus niger, Penicillium camembertii*), are inactive toward biodiesel synthesis in the described conditions.

Enzyme immobilization on Accurel could be performed by direct contact between the lipase solution and the support (Cheirsilp et al., 2008). However, it has been confirmed that ethanol pre-treatment improves the immobilization process by inducing a better penetration of the enzyme solution inside the hydrophobic Accurel and by reducing the enzyme thermodynamic activity, thus forcing the adsorption process (Foresti, & Ferreira, 2004). Enzyme adsorption with previous ethanol treatment of the support is carried out via: (i) wetting the support with, sequentially: ethanol, aqueous ethanol solution and finally with water, with intermediary filtration, or (ii) a single wetting with ethanol and then a direct contact with the enzyme solution without removing ethanol.

Accurel, due to its hydrophobic properties, should stabilise the enzyme in its open (active) conformation. Thus, it is considered as efficient for lipase immobilization.

Other synthetic polymers used for lipases immobilization comprise: hydrophobic polystyrene macroporous resin (Li & Yan, 2010), electrospun polyacrylonitrile nanofibers with higher porosity and interconnectivity compared with other nanostructured carriers (Sakai et al., 2010), polymethacrylate (Salis et al., 2009), etc. The naturally occurring materials used as carriers for lipase immobilization include: activated carbon (Moreno-Parajàn & Giraldo, 2011; Naranjo et al., 2010) and carbon cloth (Naranjo et al., 2010), celite (Ji et al., 2010; Shah & Gupta, 2007); hydrotalcite (Yagiz et al., 2007; Zeng et al., 2009), zeolites (Yagiz et al., 2007), etc. The role of the nature of the support surface on the loading and the activity, as well as on the operational stability of the immobilized enzyme has been investigated in details.

Carrier	Immobilized lipase origin	Reference
Accurel	*Candida Antarctica*	Séverac et al., 2011; Tongboriboon et al., 2010
	Candida rugosa	Salis et al., 2008 ; Tongboriboon et al., 2010
	Pseudomonas cepacia	Salis et al., 2008 ; Tongboriboon et al., 2010
	Pseudomonas sp.	Cheirsilp et al., 2008
	Pseudomonas fluorescens	Salis et al., 2008, 2009
	Pseudomonas fluorescens	Tongboriboon et al., 2010
	Mucor javanicus	Salis et al., 2008
	Penicillium roqueforti	Salis et al., 2008
	Penicillium camembertii	Salis et al., 2008
	Rhizopus oryzae	Salis et al., 2008
	Thermomyces lanuginosus	Tongboriboon et al., 2010
Activated carbon	*Candida Antarctica*	Naranjo et al., 2010
	Candida rugosa	Moreno-Parajàn & Giraldo, 2011
Celite	*Candida rugosa*	Shah & Gupta, 2007
	Pseudomonas cepacia	Shah & Gupta, 2007
	Pseudomonas aeroginosa	Ji et al., 2010
	Pseudomonas fluorescens	Shah & Gupta, 2007
Polystyrene	*Pseudomonas cepacia*	Li & Yan, 2010
Carbon cloth	*Pseudomonas cepacia*	23 Naranjo et al., 2010
Poly(acrylonitrile)	*Pseudomonas cepacia*	16 Sakai et al., 2010
Ceramics	*Pseudomonas cepacia*	Shah & Gupta, 2007
Pre-treated textile	*Candida sp.*	Chen et al., 2009; Li et al., 2010 ; Lu et al., 2007 and 2010
Hydrophilic resins	*Rhizomucor miehei*	De Paola et al., 2009
Silica	*Rhizomucor miehei*	Chen et al., 2009
	Pseudomonas fluorescens	Salis et al., 2009
Mg-Al hydrotalcites	*Saccharomyces cerevisiae*	Zeng et al., 2009
Resin D4020	*Penicillium expansum*	Li et al., 2009
Polymethacrylate	*Pseudomonas fluorescens*	Salis et al., 2009
Organosilicate	*Pseudomonas fluorescens*	Salis et al., 2009
Hydrotalcite	*Thermomyces lanuginosus*	Yagiz et al., 2007
Zeolites	*Thermomyces lanuginosus*	Yagiz et al., 2007

Table 1. Carrier used for lipases immobilization by adsorption.

3.2 Lipases immobilization by entrapment and/or encapsulation

Entrapment involves capture of the enzyme within a matrix of a polymer, although enzyme encapsulation refers to the formation of a membrane-like physical barrier around the enzyme preparation (Cao, 2005). The matrix is usually formed during the process of the immobilization. The enzyme entrapped in a gel matrix can be further encapsulated. Both processes require simple equipment and relatively inexpensive reagents. It is supposed that enzymes immobilized by entrapment and/or encapsulation are more stable than the physically adsorbed ones. At the same time the immobilized enzymes maintain their activity and stability.

Numerous materials and techniques have been used for lipases entrapment and/or encapsulation. Some of the immobilization matrices developed during the last years (2005-2011) are enumerated in Table 2.

Carrier	Immobilized lipase origin	References
κ-carrageenan	*Candida Antarctica*	Jegannathan et al., 2010
	Candida rugosa	Jegannathan et al., 2010
	Burkholderia cepacia	Jegannathan et al., 2009, 2010
	Pseudomonas fluorescens	Jegannathan et al., 2010
	Aspergillus niger	Jegannathan et al., 2010
Silica gel	*Thermomyces lanuginosus*	Khor et al., 2010
	R. miehei	Macario et al., 2009
	Pseudomonas cepacia	Noureddini et al., 2005
Celite supported sol-gel	*Candida Antarctica*	Meunier & Legge, 2010
	Lipase NS44035	Meunier & Legge, 2010
Silica aerogel	*Candida Antarctica*	Nassreddine et al., 2008
	Candida Antarctica	Orçaire et al., 2006
	Burkholderia cepacia	Orçaire et al., 2006

Table 2. Carrier used for lipases immobilization by entrapment and/or encapsulation.

For instance, a simple technique for lipase encapsulation in κ-carrageenan by co-extrusion was suggested by Jegannathan et al. (2009, 2010). Carrageenan has been selected because of its availability, biodegradability, low cost, and lack of toxicity. It was found that at optimized reaction conditions a methyl ester conversion up to 100% could be achieved in transesterification of palm oil using the liquid core encapsulated lipase PS from *Burkholderia cepacia*. The immobilized lipase was stable and retained 82% relative transesterification activity after five cycles.

Another technique for lipase immobilization by entrapment and/or encapsulation, which has received a considerable attention in recent years, is the sol-gel process. The method involves an aqueous solution of the enzyme, a catalyst (NaOH, NaF, HCl), and an inorganic-organic matrix precursor (alkoxysilane). The hydrolysis and condensation of the precursor result in an amorphous silica matrix that covers the enzyme. The method has been applied for *R. miehei* lipase encapsulation within the micellar phase of a surfactant that is self-assembled with silica (Macario et al., 2009). It has been demonstrated that the enzyme preserves its mobility and activity. More over, because of the activation of the enzyme catalytic centre by the hydrophobic groups of the surfactant, the immobilized lipase was more active than its free form. In addition, the obtained ordered mesoporous structure improved the stability of the enzyme and decreased the rate of leaching.

Comprehensive characterization of sol–gel immobilized lipase has been performed by Noureddini et al. (2005). Lipase PS was entrapped within a sol–gel polymer matrix, prepared by polycondensation of hydrolyzed tetramethoxysilane and iso-butyltrimethoxysilane. The immobilized lipase was stable and more active than the free lipase toward the transesterification of soybean oil.

Various supports could be used to improve the stability of the entrapped/encapsulated enzymes. Celite supported lipase sol-gels were investigated aiming such problems as activity, stability and reusability of the enzyme (Meunier & Legge, 2010). The three types of Celite considered (R633, R632, and R647) were compared to unsupported lipase sol–gels. It has been established that sol–gel immobilized lipase supported on Celite R632 allowed achieving an average conversion of 60% per gram of material for 6 h, and exhibited an average initial lipase activity comparable to that of the unsupported sol–gel formulation.

Orçaire et al. (2006) report a technique for encapsulation of *Candida Antarctica* and *Burkholderia cepacia* lipases in silica aerogels reinforced with silica quartz fibre felt and dried by the CO_2 supercritical technique. The aerogel encapsulation permits maintaining the enzymes in a dispersion state similar to the dispersion prevailing in an aqueous solution, even in organic media, while agglomeration of the lipase occurs if it is used directly in the organic solvent. At present, sol–gel enzyme entrapment/encapsulation is considered to be the most successful immobilization technique for lipase immobilization.

3.3 Lipases immobilization by covalent attachment

Covalent attachment is a result of a chemical reaction between the active amino acid residues outside the active catalytic and binding site of the enzyme, and the active functionalities of the carrier (Cao, 2005). Although drastic and complicated, and strongly affected by the carriers' properties, covalent attachment is the most efficient technique for enzyme immobilization. Some carriers used for covalent lipase immobilization are displayed in Table 3.

Carrier	Immobilized lipase origin	References
Olive pomace	*Thermomyces lanuginosus*	Yücel, 2011
Resins	*Thermomyces lanuginosus*	Mendes et al., 2011
	Pseudomonas fluorescens	Mendes et al., 2011
Polymers	*Thermomyces lanuginosus*	Dizge et al., 2008, 2009a, 2009b
Polyurethane foam	*Thermomyces lanuginosus*	Dizge & Keskinler, 2008
Nb_2O_5 and SiO_2-PVA	*Burkholderia cepacia*	Da Rys et al., 2010
Chitosan	*Candida rugosa*	Shao et al., 2008
		Ting et al., 2008
Lewatit	*Thermomyces lanuginosus*	Rodrigues et al., 2010
Silica	*Rhizopus orizae+Candida rugosa*	Lee and al., 2008
	Enterobacter aerogenes	Kumari et al., 2009
Magnetic nanostructures	*Candida rugosa*	Dussan et al., 2007, 2010
	Thermomyces lanuginosus	Xie & Ma, 2010

Table 3. Carrier used for lipases immobilization by covalent attachment

Yücel (2011) reports a method for *Thermomyces lanuginosus* lipase covalent binding on polyglutaraldehyde-activated olive pomace powder. The technique is cost effective, because of the low price of the support and because of the strong covalent bond formed, leading to enzyme stabilization without loss of activity, allowing the multiple reuse of the enzyme. Immobilized lipase was stable for 10 batches of pomace oil transesterification retaining more than 80% residual activity.

Among the other naturally occurring materials, chitosan is considered as appropriate for enzyme binding. Its membrane forming and adhesion ability, high mechanical strength and facility of forming insoluble in water thermally and chemically inert films make it suitable for lipase immobilization. For instance, *Candida rugosa* type VII lipase was fixed onto chitosan beads using a binary method consisting in the follows: (i) lipase immobilization onto the hydroxyl groups of chitosan by activation with 1-ethyl-3-(3-dimethylaminopropyl) carbodiimide hydrochloride; (ii) immobilization of additional lipase molecules through their amino groups to chitosan by cross-linking with glutaraldehyde. The immobilized enzyme has been used to catalyse the hydrolysis of soybean oil. Then, the feedstock containing free fatty acids, mono-, di- and triglycerides was esterified with methanol in the presence of an acid catalyst to produce biodiesel. It has been demonstrated that the enzymatic/acid-catalyzed hybrid process uses milder reaction conditions and allows avoiding the inactivation of the immobilized enzyme by polar compounds and increase biodiesel yields.

A new method for the synthesis of hydrophobic microporous matrices for enzyme immobilization, namely styrene–divinylbenzene–polyglutaraldehyde and poly(styrene-divinylbenzene)-polyglutaraldehyde copolymers, applying High Internal Phase Emulsions (HIPE) technique has been developed by Dizge et al. (2008, 2009a, 2009b). *Thermomyces lanuginosus* lipase was successfully attached to the support by covalent binding. According to the authors, the copolymers could be prepared in a short time and in large amounts and shapes. The immobilization efficiency, defined as the ratio of the activity of the immobilized enzyme to the activity of the free enzyme was found to vary from 80% to 89%. The immobilized enzyme retained its activity during 10-15 repeated batch reactions.

Promising results in terms of enzyme thermal and operational stability improvement have been obtained using as supports for lipases immobilization silanized Nb_2O_5 and SiO_2-PVA (Da Rys et al., 2010), and glutaraldehyde or ethanolamine activated silica gels (Lee and al., 2008; Kumari et al., 2009). However, the stability of the immobilized enzyme depends not only on the chemical/physical nature of the carrier, but also on the binding mode, the binding number, and the position of the binding on the enzyme surface (Cao, 2005), among other. Mendes et al. (2011) and Rodrigues et al. (2010) demonstrate that the multipoint covalent attachment of *Thermomyces lanuginosus* lipase on Toyopearl AF-amino-650M resin and on aldehyde-Lewatit is an efficient strategy for enzyme stabilization. It has been demonstrated that *Thermomyces lanuginosus* lipase immobilized on glyoxyl-resin is between 27 and 31 times more stable than the soluble lipase (Mendes et al., 2011).

Another important issue provided by enzyme immobilization concerns the localization of the interaction in the zone where the maximum concentration of reagents is present and/or at the interface between the immiscible heterogeneous phases, regarding lipases. For this purpose, lipases (from *Candida rugosa* and *Thermomyces lanuginosus*) have been immobilized on magnetic nanostructures (Fe_3O_4) and localized by application of a magnetic field. In addition, the method favours the simple and fast separation of the enzyme from the reaction mixture. Thus, it allows the intensification of the process and the reduction of the production costs.

3.4 Cells immobilization

The technological and economic advantages of immobilized cells over immobilized enzymes are well known (D'Souza, 1982): higher operational stability, higher yields of enzyme activity after immobilization, greater resistance to environmental perturbations, greater potential for multistep processes, and lower effective enzyme cost (enzyme purification and extraction are avoided). Despite of these benefits, only few investigations on the use of immobilized cells in biodiesel synthesis are reported until now (Fukuda et al., 2009; Hama et al., 2006, 2007; Li et al., 2007; Oda et al., 2005; Tamalampudi et al., 2008). The research efforts were focused on the immobilization no more than of *Rhysopus oryzae* within porous biomass support particles. The fixation was achieved spontaneously during batch cultivation. The applied technique (Atkinson et al., 1979) offers numerous advantages over other methods: the particles are reusable and mechanically resistant; additional reagents, aseptic handling of particles, and preproduction of cells are not necessary. It has been demonstrated that *Rhysopus oryzae* cells immobilized within biomass support particles can be used as low cost catalyst for biodiesel production.

4. Conclusion

Enzymatic approach to biodiesel production offers several advantages over the chemical catalysis currently applied. It is more efficient because of the enzyme specificity and selectivity, involves less energy consumption because of the mild reaction conditions, and is environmentally friendly because of the limited release of side products or wastes. Catalyst immobilization presents a number of additional benefits, such as repeated use of the enzymes, enhancement of their thermal and operational stability, localization of the interaction, effective control of the reaction parameters, etc., thus reducing the production cost and making the enzyme biodiesel synthesis an attractive alternative to other technologies. The present review provides an overview on the techniques applied for lipases immobilization.

5. References

Adinarayana, K.; Bapi Raju, K. V. V. S. N.; Zargar, M. I. ; Devi, R. B.; Lakshmi, P. J. & Ellaiah, P. (2004). Optimization of process parameters for production of lipase in solid-state fermentation by newly isolated *Aspergillus* species, *Ind. J. Biotechnol.*, 3, 65-69

Ahn, K.; Ye, S.; Chun, W.; Rah, H. & Kim, S. (2010). Yield and component distribution of biodiesel by methanolysis of soybean oil with lipase-immobilized mesoporous silica. *Microporous and Mesoporous Materials* doi:10.1016/j.micromeso.2010.11.008 (in press)

Akoh, C.; Chang, Shu-Wei; Lee, Guan-Chiun & Shaw, Jei-Fu (2007). Enzymatic approach to biodiesel production. *J. Agric. Food Chem.* 55, 8995-9005

ASTM D6751: American Society for Testing and Materials standard specification for biodiesel fuel (B 100) blend stock for distillate fuels

Atkinson, B.; Black, G. M.; Lewis, P. J. S. & Pinches, A. (1979). Biological particles of given size, shape, and density for use in biological reactors. *Biotechnol. Bioeng.*, 21, 193-200

Bajaj, A.; Lohan, P.; Jha, P. & Mehrotra, R. (2010). Biodiesel production through lipase catalyzed transesterification: An overview. *J. Mol. Catal. B: Enzymatic*, 62, 9-14

Bisen, P.; Sanodiya, B.; Thakur, G.; Baghel, R. & Prasad, G. B. K. S. (2010). Biodiesel production with special emphasis on lipase-catalyzed transesterification. *Biotechnol. Lett.*, 32, 1019–1030

Cao, L. (2005). *Carrier-bound immobilized enzymes. Principles, application and design.* Wiley-VCH, ISBN-10: 3-527-31232-3, Weinheim

Chahinian, H.; Vanot, G.; Ibrik, A.; Rugani, N.; Sarda, L. & Comeau, L. C. (2000). Production of extracellular lipases by *Penicillium cyclopium* purification and characterization of a partial acylglycerol lipase. *Biosci. Biotechnol. Biochem.*, 64, 215-222

Cheirsilp, B.; H-Kittikuna, A. & Limkatanyu, S. (2008). Impact of transesterification mechanisms on the kinetic modeling of biodiesel production by immobilized lipase. *Biochem. Eng. J.*, 42, 261–269

Chen, Y.; Xiao, B.; Chang, J.; Fu, Y.; Lv, P. & Wang, X. (2009). Synthesis of biodiesel from waste cooking oil using immobilized lipase in fixed bed reactor. *Energy Conversion and Management*, 50, 668–673

Da Rys, P.; Silva, G.; Mendes, A.; Santos, J. & Castro, H. (2010). Evaluation of the catalytic properties of *Burkholderia cepacia* lipase immobilized on non-commercial matrices to be used in biodiesel synthesis from different feedstocks. *Bioresource Technology*, 101, 5508–5516

De Paola, M. G.; Ricca, E.; Calabrò, V.; Curcio, S. & Iorio, G. (2009). Factor analysis of transesterification reaction of waste oil for biodiesel production. *Bioresource Technology*, 100, 5126–513

Demirbas, A. (2009). Progress and recent trends in biodiesel fuels. *Energy Conversion and Management*, 50, 14–34

Dizge, N. & Keskinler, B. (2008). Enzymatic production of biodiesel from canola oil using immobilized lipase. *Biomass and Bioenergy*, 32, 1274–1278

Dizge, N.; Keskinler, B. & Tanriseven, A. (2008). Covalent attachment of microbial lipase onto microporous styrene-divinylbenzene copolymer by means of poly glutaraldehyde. *Colloids and Surfaces B: Biointerfaces*, 66, 34–38

Dizge, N.; Keskinler, B. & Tanriseven, A. (2009a). Biodiesel production from canola oil by using lipase immobilized onto hydrophobic microporous styrene-divinylbenzene copolymer. *Biochem. Eng. J.*, 44, 220–225

Dizge, N.; Aydiner, C.; Imer, D.; Bayramoglu, M.; Tanriseven, A. & Keskinler, B. (2009b). Biodiesel production from sunflower, soybean, and waste cooking oils by transesterification using lipase immobilized onto a novel microporous polymer. *Bioresource Technology*, 100, 1983–1991

D'Souza, S. F. (1989). Immobilized cells: techniques and applications. *Indian J. Microbiol.*, 29, 83–117

Dussán, K.; Giraldo, O. & Cardona, C. (2007). Application of magnetic nanostructures in biotechnological processes: Biodiesel production using lipase immobilized on magnetic carriers. *Proceedings of European Congress of Chemical Engineering* (ECCE-6), Copenhagen, 16-20 September 2007

Dussan, K.; Cardona, C.; Giraldo, O.; Gutiérrez, L. & Pérez, V. (2010). Analysis of a reactive extraction process for biodiesel production using a lipase immobilized on magnetic nanostructures. *Bioresource Technology*, 101, 9542–9549

Fjerbaek, L.; Christensen, K. & Norddahl, B. (2009). A Review of the Current State of Biodiesel Production Using Enzymatic Transesterification. *Biotechnol Bioeng.*, 102, 5, 1298-1315

Foresti, M. L. & Ferreira M. L. (2004). Ethanol pretreatment effect and particle diameter issues on the adsorption of *Candida rugosa* lipase onto polypropylene powder. *Appl. Surf. Sci.*, 238, 86–90

Fukuda, H.; Kondo, A. & Noda, H. (2001). Biodiesel fuel production by transesterification of oils. *J. Bioscience Bioeng.*, 5, 405–416

Fukuda, H.; Kondo, A. & Tamalampudi, S. (2009). Bioenergy: Sustainable fuels from biomass by yeast and fungal whole-cell biocatalysts. *Biochem. Eng. J.*, 44, 2–12

Ghaly, A. E.; Dave, D.; Brooks, M. S. & Budge, S. (2010). Production of biodiesel by enzymatic transesterification: Review. *Am. J. Biochem. Biotechnol.*, 6, 54-76

Guncheva, M. & Zhiryakova, D. (2011). Catalytic properties and potential applications of Bacillus lipases. *J. Mol. Catalysis B: Enzymatic*, 68, 1-21

Hama, S.; Tamalampudi, S.; Fukumizu, T.; Miura, K.; Yamaji, H.; Kondo, A. & Fukuda, H. (2006). Lipase localization in *Rhizopus oryzae* cells immobilized within biomass support particles for use as whole-cell biocatalysts in biodiesel-fuel production. *J. Bioscience Bioeng.*, 101, 328-333

Hama, S.; Yamaji, H.; Fukumizu, T.; Numata, T.; Tamalampudi, S.; Kondo, A. Noda, H. & Fukuda, H. (2007). Biodiesel-fuel production in a packed-bed reactor using lipase-producing *Rhizopus oryzae* cells immobilized within biomass support particles. *Biochem. Eng. J.*, 34, 273-278

Hasan, F.; Shah, A. & Hameed, A. (2009). Methods for detection and characterization of lipases: A comprehensive review. *Biotechnol. Adv.*, 27, 782-798

Helwani, Z.; Othman, M. R.; Aziz, N.; Fernando, W. J. N. & Kim, J. (2009). Technologies for production of biodiesel focusing on green catalytic techniques: A review. *Fuel Processing Technology*, 90, 1502–1514

Hiol, A.; Jonzo, M. D.; Rugani, N.; Druet, D.; Sarda, L. & Comeau, L. C. (2000). Purification and characterization of an extracellular lipase from a thermophilic *Rhizopus oryzae* strain isolated from palm fruit. *Enzyme Microb. Technol.*, 26, 421–30

Jegannathan, K.; Jun-Yee, L.; Chan, E. & Ravindra, P. (2009). Design an immobilized lipase enzyme for biodiesel production. *J. Renewable and Sustainable Energy*, 1, 063101-1 - 063101-8

Jegannathan K. Abang, S. (2008). Production of biodiesel using immobilized lipase-A critical review. *Critical Reviews in Biotechnology*, 28, 253–264

Jegannathan, K.; Jun-Yee, L.; Chan, E. & Ravindra, P. (2010). Production of biodiesel from palm oil using liquid core lipase encapsulated in κ-carrageenan. *Fuel*, 89, 2272–2277

Ji, Q.; Xiao, S.; He, B. & Liu, X. (2010). Purification and characterization of an organic solvent-tolerant lipase from *Pseudomonas aeruginosa* LX1 and its application for biodiesel production. *J. Mol. Catal. B: Enzymatic*, 66, 264–269

Karanam, S. K. & Medicherla, N. R. (2008). Enhanced lipase production by mutation induced *Aspergillus japonicus*. *African J. Biotechnol.*, 7, 2064-2067

Kashmiri, M. A.; Adnan, A.; Butt, B. W. (2006). Production, purification and partial characterization of lipase from *Trichoderma viride*. *African J. Biotechnol.*, 5, 878-882

Khor, G.; Sim, J.; Kamaruddin, A. & Uzir, M. (2010). Thermodynamics and inhibition studies of lipozyme TL IM in biodiesel production via enzymatic transesterification. *Bioresource Technology*, 101, 6558–6561

Kumari, A.; Mahapatra, P.; Garlapati, V. & Banerjee, R. (2009). Enzymatic transesterification of Jatropha oil. *Biotechnology for Biofuels*, 2:1, doi:10.1186/1754-6834-2-1

Lee Jong Ho; Dong Hwan Lee; Jung Soo Lim; Byung-Hwan Um; Chulhwan Park; Seong Woo Kang & Seung Wook Kim (2008). Optimization of the process for biodiesel production using a mixture of immobilized *Rhizopus oryzae* and *Candida rugosa* lipases. *J. Microbiol. Biotechnol.*, 18, 1927–1931

Li, N.; Zong, M.; Wu, H. (2009). Highly efficient transformation of waste oil to biodiesel by immobilized lipase from *Penicillium expansum*. *Process Biochemistry*, 44, 685–688

Li, Q. & Yan, Y. (2010). Production of biodiesel catalyzed by immobilized *Pseudomonas cepacia* lipase from Sapium sebiferum oil in micro-aqueous phase. *Applied Energy*, 87, 3148–3154

Li, W.; Du, W. & Liu, D. (2007). Rhizopus oryzae IFO 4697 whole cell catalyzed methanolysis of crude and acidified rapeseed oils for biodiesel production in tert-butanol system. *Process Biochemistry*, 42, 1481–1485

Li, Z.; Deng, L.; Lu, J.; Guo, X.; Yang, Z. & Tan, T. (2010). Enzymatic synthesis of fatty acid methyl esters from crude rice bran oil with immobilized *Candida sp.* 99-125. *Chinese Journal of Chemical Engineering*, 18, 870–875

Lima, V. M. G.; Krieger, N.; Sarquis, M. I. M.; Mitchell, D. A.; Ramos, L. P. & Fontana, J. D. (2003). Effect of nitrogen and carbon sources on lipase production by *Penicillium aurantiogriseum*. *Food Technol. Biotechnol.*, 41, 105–110

Lu, J.; Nie, K.; Xie, F.; Wang, F. & Tan, T. (2007). Enzymatic synthesis of fatty acid methyl esters from lard with immobilized *Candida sp.* 99-125. *Process Biochemistry*, 42, 1367–1370

Lu, J.; Deng, L.; Zhao, R.; Zhang, R.; Wang, F. & Tan, T. (2010). Pretreatment of immobilized *Candida sp.* 99-125 lipase to improve its methanol tolerance for biodiesel production. *J. Mol. Catal. B: Enzymatic*, 62, 15–18

Macario, A.; Moliner, M; Corma, A. & Giordano, G. (2009). Increasing stability and productivity of lipase enzyme by encapsulation in a porous organic–inorganic system. *Microporous and Mesoporous Materials*, 118, 334–340

Man Xi Ao, Sini Mathew & Obbard, J. (2009). Biodiesel fuel production via transesterification of oils using lipase biocatalyst. *GCB Bioenergy*, 1, 115–125

Marchetti, J. M.; Miguel, V. U. & Errazu, A.F. (2007). Possible methods for biodiesel production. *Renewable and Sustainable Energy Reviews*, 11, 1300–1311

Mendes, A.; Giordano, R. C.; Giordano, R. L. C. & Castro, H. (2011). Immobilization and stabilization of microbial lipases by multipoint covalent attachment on aldehyde-resin affinity: Application of the biocatalysts in biodiesel synthesis. *J. Mol. Catal. B: Enzymatic*, 68, 109–115

Meunier, S. & Legge, R. (2010). Evaluation of diatomaceous earth as a support for sol–gel immobilized lipase for transesterification. *J. Mol. Catal. B: Enzymatic*, 62, 54–58

Moreno-Parajàn, J. C. & Giraldo, L. (2011). Study of immobilized candida rugosa lipase for biodiesel fuel production from palm oil by flow microcalorimetry. *Arabian Journal of Chemistry*, 4, 55–62

Naranjo, J.; Córdoba, A.; Giraldo, L.; García, V. & Moreno-Parajàn, J. C. (2010). Lipase supported on granular activated carbon and activated carbon cloth as a catalyst in the synthesis of biodiesel fuel. *J. Mol. Catal. B: Enzymatic*, 66, 166–171

Nassreddine, S.; Karout, A.; Christ, M. & Pierre, A. (2008). Transesterification of a vegetal oil with methanol catalyzed by a silica fibre reinforced aerogel encapsulated lipase. *Applied Catalysis A: General*, 344, 70–77

Oda, M.; Kaieda, M.; Hama, S.; Yamaji, H.; Kondo, A.; Izumoto, E. & Fukuda, H. (2005). Facilitatory effect of immobilized lipase-producing *Rhizopus oryzae* cells on acyl migration in biodiesel-fuel production. *Biochem. Eng. J.*, 23, 45–51

Orçaire, O.; Buisson, P. & Pierre, A. (2006). Application of silica aerogel encapsulated lipases in the synthesis of biodiesel by transesterification reactions. *J. Mol. Catal. B: Enzymatic*, 42, 106–113

Noureddini, H.; Gao, X. & Philkana, R. S. (2005). Immobilized Pseudomonas cepacia lipase for biodiesel fuel production from soybean oil. *Bioresource Technology*, 96, 769–777

Panalotov, I. & Verger, R. (2000). *Physical Chemistry of Biological Interfaces*. Marcel Dekker Inc., New York

Powel, L. W. (1996). In *Industrial Enzymology*, T. Godfrey and S. West, Editors, 2nd ed. Stockton Press, NY

Rajesh, E. M.; Arthe, R.; Rajendran, R.; Balakumar, C.; Pradeepa, N. & Anitha, S. (2010). Investigation of lipase production by *Trichoderma reesei* and optimization of production parameters. *EJEAFChe*, 9, 1177-1189

Ranganathan, S.; Narasimhan, S. & Muthukumar, K. (2008). An overview of enzymatic production of biodiesel. *Bioresource Technology*, 99, 3975–3981

Robles-Medina, A.; González-Moreno, P.A.; Esteban-Cerdán, L. & Molina-Grima, E. (2009). Biocatalysis: Towards ever greener biodiesel production. *Biotechnology Advances*, 27, 398–408

Rodrigues, R. & Fernandez-Lafuente, F. (2010). Lipase from Rhizomucor miehei as an industrial biocatalyst in chemical process. *J. Mol. Catal. B: Enzymatic*, 64, 1–22

Rodrigues, R.; Pessela, B.; Volpato, G.; Fernandez-Lafuente, R.; Guisan, J. & Ayub, M. (2010). Two step ethanolysis: A simple and efficient way to improve the enzymatic biodiesel synthesis catalyzed by an immobilized–stabilized lipase from *Thermomyces lanuginosus*. *Process Biochemistry*, 45, 1268–1273

Sakai, S.; Yuping Liu, Y.; Yamaguchi, T.; Watanabe, R.; Kawabe, M. & Kawakami, K. (2010). Production of butyl-biodiesel using lipase physically-adsorbed onto electrospun polyacrylonitrile fibers. *Bioresource Technology*, 101, 7344–7349

Salis, A.; Pinna, M.; Monduzzi, M. & Solinas, V. (2008). Comparison among immobilised lipases on macroporous polypropylene toward biodiesel synthesis. *J. Mol. Catal. B: Enzymatic*, 54, 19–26

Salis, A.; Bhattacharyya, M.; Monduzzi, M. & Solinas, V. (2009). Role of the support surface on the loading and the activity of *Pseudomonas fluorescens* lipase used for biodiesel synthesis. *J. Mol. Catal. B: Enzymatic*, 57, 262–269

Semwal, S.; Arora, A.; Badoni, R. & Tuli. D. (2011). Biodiesel production using heterogeneous catalysts. *Bioresour. Technol.*, 102, 2151-2161

Séverac, E.; Galy, O.; Turon, F.; Pantele, C.; Condoret, J-S.; Monsan, P. & Marty, A. (2011). Selection of CalB immobilization method to be used in continuous oil transesterification: Analysis of the economical impact. *Enzyme and Microbial Technology*, 48, 61–70

Shah, S. & Gupta, M. (2007). Lipase catalyzed preparation of biodiesel from Jatropha oil in a solvent free system. *Process Biochemistry*, 42, 409–414

Shao, P.; Meng, X.; He, J. & Sun, P. (2008). Analysis of immobilized *Candida rugosa* lipase catalyzed preparation of biodiesel from rapeseed soapstock. *Food and Bioproducts Processing*, 8 6, 283–289

Shukla, P. & Gupta, K. (2007). Ecological screening for lipolytic molds and process optimization for lipase production from *Rhizopus oryzae* KG-5. *J Appl. Sci. Environ. Sanit.*, 2, 35–42

Tamalampudi, S.; Talukder, M.; Hama, S.; Numata, T.; Kondo, A. & Fukuda, H. (2008). Enzymatic production of biodiesel from Jatropha oil: A comparative study of immobilized-whole cell and commercial lipases as a biocatalyst. *Biochem. Eng. J.*, 39, 185–189

Tan, T.; Lu, J.; Nie, K.; Deng, L. & Wang, F. (2010). Biodiesel production with immobilized lipase: A review. *Biotechnology Advances*, 28, 628–634

Ting, W.; Huang, C.; Giridhar, N. & Wu, W. (2008). An enzymatic/acid-catalyzed hybrid process for biodiesel production from soybean oil. J. *Chinese Institute of Chemical Engineers*, 39, 203–210

Tongboriboon, K.; Cheirsilp, P. & H-Kittikun, A. (2010). Mixed lipases for efficient enzymatic synthesis of biodiesel from used palm oil and ethanol in a solvent-free system. *J. Mol. Catal. B: Enzymatic*, 67, 52–59

Treichel, H.; Oliveira, D.; Mazutti, M.; Di Luccio, M. & Oliveira, J. (2010). A Review on Microbial Lipases Production. *Food Bioprocess Technol.*, 3, 182–196

Vardanega, R.; Remonatto, D.; Arbter, F.; Polloni, A.; Rigo, E.; Ninow, L. N.; Treichel, H.; Oliveira, D. & Luccio, M. (2010). A Systematic study on extraction of lipase obtained by solid-state fermentation of soybean meal by a newly isolated strain of *Penicillium* sp. *Food Process. Technol.*, 3, 461-465

Xie, W. & Ma, N. (2010). Enzymatic transesterification of soybean oil by using immobilized lipase on magnetic nano-particles. *Biomass and Bioenergy*, 34, 890-896

Yagiz, F; Kazan, D. & Nilgun Akin, A. (2007). Biodiesel production from waste oils by using lipase immobilized on hydrotalcite and zeolites. *Chem. Eng. J.*, 134, 262–267

Yücel, Y. (2011). Biodiesel production from pomace oil by using lipase immobilized onto olive pomace. *Bioresource Technology*, 102, 3977–3980

Zeng, H.; Liao, K.; Deng, X.; Jiang, H. & Zhang, F. (2009). Characterization of the lipase immobilized on Mg–Al hydrotalcite for biodiesel. *Process Biochemistry*, 44, 791–798

Permissions

The contributors of this book come from diverse backgrounds, making this book a truly international effort. This book will bring forth new frontiers with its revolutionizing research information and detailed analysis of the nascent developments around the world.

We would like to thank Margarita Stoytcheva and Gisela Montero, for lending their expertise to make the book truly unique. They have played a crucial role in the development of this book. Without their invaluable contribution this book wouldn't have been possible. They have made vital efforts to compile up to date information on the varied aspects of this subject to make this book a valuable addition to the collection of many professionals and students.

This book was conceptualized with the vision of imparting up-to-date information and advanced data in this field. To ensure the same, a matchless editorial board was set up. Every individual on the board went through rigorous rounds of assessment to prove their worth. After which they invested a large part of their time researching and compiling the most relevant data for our readers. Conferences and sessions were held from time to time between the editorial board and the contributing authors to present the data in the most comprehensible form. The editorial team has worked tirelessly to provide valuable and valid information to help people across the globe.

Every chapter published in this book has been scrutinized by our experts. Their significance has been extensively debated. The topics covered herein carry significant findings which will fuel the growth of the discipline. They may even be implemented as practical applications or may be referred to as a beginning point for another development. Chapters in this book were first published by InTech; hereby published with permission under the Creative Commons Attribution License or equivalent.

The editorial board has been involved in producing this book since its inception. They have spent rigorous hours researching and exploring the diverse topics which have resulted in the successful publishing of this book. They have passed on their knowledge of decades through this book. To expedite this challenging task, the publisher supported the team at every step. A small team of assistant editors was also appointed to further simplify the editing procedure and attain best results for the readers.

Our editorial team has been hand-picked from every corner of the world. Their multi-ethnicity adds dynamic inputs to the discussions which result in innovative outcomes. These outcomes are then further discussed with the researchers and contributors who give their valuable feedback and opinion regarding the same. The feedback is then collaborated with the researches and they are edited in a comprehensive manner to aid the understanding of the subject.

Apart from the editorial board, the designing team has also invested a significant amount of their time in understanding the subject and creating the most relevant covers. They scrutinized every image to scout for the most suitable representation of the subject and create an appropriate cover for the book.

The publishing team has been involved in this book since its early stages. They were actively engaged in every process, be it collecting the data, connecting with the contributors or procuring relevant information. The team has been an ardent support to the editorial, designing and production team. Their endless efforts to recruit the best for this project, has resulted in the accomplishment of this book. They are a veteran in the field of academics and their pool of knowledge is as vast as their experience in printing. Their expertise and guidance has proved useful at every step. Their uncompromising quality standards have made this book an exceptional effort. Their encouragement from time to time has been an inspiration for everyone.

The publisher and the editorial board hope that this book will prove to be a valuable piece of knowledge for researchers, students, practitioners and scholars across the globe.

List of Contributors

Kandukalpatti Chinnaraj Velappan and Nagarajan Vedaraman
Chemical Engineering Department, Central Leather Research Institute, Council of Scientific and Industrial Research, Adyar, Chennai, India

Paula Mazo, Gloria Restrepo and Luis Rios
Universidad de Antioquia, Grupo Procesos Fisicoquímicos Aplicados, Colombia

Asnida Yanti Ani, Mohd Azlan Mohd Ishak and Khudzir Ismail
Fossil & Biomass Energy Research Group, Fuel Combustion Research Laboratory, Universiti Teknologi MARA, Perlis, Malaysia

Somkiat Ngamprasertsith and Ruengwit Sawangkeaw
Fuels Research Center, Department of Chemical Technology, Faculty of Science, Chulalongkorn University, Thailand
Center for Petroleum, Petrochemicals and Advance Materials, Chulalongkorn University, Thailand

G.B. Shinde
Department of Chemical Engineering, Sir Visvesvaraya Institute of Technology, Nashik, M.S., India

V.S. Sapkal
Sant Tukadoji Maharaj Nagpur University, Nagpur, M.S., India

R.S. Sapkal
University Department of Chemical Technology, Sant Gadgebaba Amravati University, Amravati, M.S., India

N.B. Raut
Faculty of Engineering, Sohar University, Sultanate of Oman, Oman

Amrit Goswami
Chemical Science Block, CSIR- North-East Institute of Science & Technology, (A Constituent Establishment of Council of Scientific and Industrial Research, Govt. of India), Jorhat-785006, Assam, India

Feng Guo and Zhen Fang
Chinese Academy of Sciences, Biomass Group, Xishuangbanna Tropical Botanical Garden, China

Marcio Jose da Silva, Abiney Lemos Cardoso, Fernanda de Lima Menezes and Aline Mendes de Andrade
Federal University of Viçosa/Chemistry Department, Brazil

Manuel Gonzalo Hernandez Terrones
Federal University of Uberlândia/Chemistry Institute, Brazil

Ana Aurelia Chirvase and Nicoleta Radu
The National Institute for Research & Development in Chemistry and Petrochemistry, Romania

Luminita Tcacenco
The National Institute for Research & Development in Biological Sciences, Romania

Irina Lupescu
The National Institute for Research & Development in Pharmaceutical Chemistry, Romania

Carlos Vera, Mariana Busto, Juan Yori, Gerardo Torres, Debora Manuale, Sergio Canavese and Jorge Sepúlveda
INCAPE (FIQ, Universidad Nacional del Litoral-CONICET), Argentina

Margarita Stoytcheva, Gisela Montero, Lydia Toscano and Benjamin Valdez
Universidad Autónoma de Baja California, Instituto de Ingeniería, México

Velizar Gochev
Plovdiv University "Paisii Hilendarski, Bulgaria